Gravity: what we still don't know

Many physicists think general relativity is a large-scale approximation, and not the last word

Jonathan Kerr

The present view of gravity is now being questioned: it doesn't fit with quantum theory, other theories predict results like gravitational waves. And why does the apple <u>start</u> to fall? Surprisingly perhaps, there's no consensus, and it's unexplained

Gb

Published by Gordon Books

April Cottage, 9 The Green, Clophill, Bedfordshire MK45 4AD

The publication of this book establishes the author's origination of some of the ideas in it (and as intellectual property), as it precedes any other publication of them, except certain self-contained areas by the author, in peer reviewed journals and in Book I of this series, 'The Unsolved Puzzle'. The set of ideas in the later part of this book, in Part 5 and onwards, Planck Scale Gravity theory (PSG, part of the same overall theory as dimensional quantum mechanics, or DQM), was first discovered in 1998 and developed until 2022.

21st May 2022

ISBN
Paperback: 978-0-9564222-3-1
E-book: 978-0-9564222-8-6

Once in motion, the present view of gravity describes the apple's trajectory beautifully. But why it starts to move is a mystery in physics, and there's no consensus on how to explain it

----

With thanks above all to Nigel Lesmoir-Gordon, for help with the book, for making the documentary, and for his steadfast, cheerful encouragement over many years. Also with thanks to Jenny Lesmoir-Gordon, for helpful input and general support throughout, and to their son Gabriel Lesmoir-Gordon for work on the artwork. And thanks to Paul 'Joey' Wheeler, for his advice and help with all kinds of things. And thanks to the people who contributed valuable discussions, in particular Carlo Rovelli and Neil Turok, for long conversations on quantum mechanics and other areas relating to the wider theory. And thanks to Jill Pliskin, for all the times of mutual support over many years, during the work on the book.

The chapters in this book are short, averaging less than two pages. So the index is long, and is at the end of the book, with the alphabetical index and reference section.

Some links relating to Book I:

New theory of quantum mechanics shows matter is not in the eye of the observer
Sunday Telegraph: https://archive.ph/CxFoI
"Quantum mechanics' greatest puzzle 'solved' in a Surrey cottage" [Print version, Pressreader]

Paper: *An interactions-based interpretation for quantum mechanics*
https://doi.org/10.4006/0836-1398-33.1.1

Documentary *'The Interactions Avenue', partly a conversation between the author and Carlo Rovelli:* https://www.interactionsavenue.com

Introduction

Physicists take different views on gravity, but the majority think our ideas on it are incomplete. A lot of people - that is, people in mainstream physics - say we still don't understand it. We have a theory, general relativity, that gives a very accurate description of the effects of gravity. But it doesn't cover the causes half as well, and vital questions remain.

Over the last few decades there has been a growing realisation that attempts to connect our two main theories are simply not working. General relativity and quantum mechanics contradict each other. It used to look certain that they would be fused into quantum gravity - the drive to make that happen has been the central project of physics since the 1980s. String theory used to look like it would do it, but not any more. A lot of physicists think a major adjustment is needed, but only recently, even ten years ago, people thought it might be a minor tweak here or there.

This means our whole view of the physical world doesn't hang together. And when you ask well-respected physicists which of these two theories will need changing or replacing, if they say anything, they always say general relativity. Quantum mechanics is unassailable, but general relativity behaves as a large-scale approximation: calculations using it at a small scale will sometimes give nonsense. This was a taboo, but now people are saying it out loud. General relativity is extremely good within a limited domain, as Newton's theory was. Many quietly think it will have to be replaced in the same way.

So although special relativity is above suspicion, nowadays the gravity theory isn't. Raphael Bousso, who is Professor at the Berkeley Centre for Theoretical Physics, University of California, is a highly regarded up and coming physicist. For that reason, he was asked to write a piece for edge.org. He says:

"General Relativity is only a classical theory. It rests on a demonstrably false premise: that position and momentum can be known simultaneously. This may be a good approximation for apples, planets, and galaxies: large objects, for which gravitational interactions tend to be much more important than for the particles of the quantum world. But as a matter of principle, the theory is wrong. The seed is there. General Relativity cannot be the final word; it can only be an approximation to a more general Quantum Theory of Gravity.

Sabine Hossenfelder, another well-known and respected physicist, said *'Most physicists, me included, think general relativity is only an approximation to a better theory'*. And elsewhere: *'But we already know that it cannot ultimately be the correct theory for space and time. It is an approximation that works in many circumstances, but fails in others.'*

It used to be a risk to one's career to say that. People still like to get to a safe position first: if you haven't heard what you'll read here, that's why. But the taboos I'll break have already been broken by honest people such as Bousso, Hossenfelder and others. The tide is turning, and the recent view of the old ideas is out - there's a fresh attitude in the 21st century, and past restrictions are fading fast.

The result is what I'll show you in this book, which is that a new, unexpected landscape appears. Everything looks different. Gravity is more than a force that pulls matter together, it's fundamental to the universe: it's connected to mass, energy, inertia, and other things we haven't explained, or have only half explained. Gravity is likely to be the key to the next paradigm, because it happens to be a large part of what's wrong with this one. A growing number have started to think this is the case.

And we find ourselves with a list of fascinating clues that were overlooked, because people used to think we already had the answer. In this book I'll go through these clues, as we now need to look at them with fresh eyes - and in relation to new data coming in, such as from black holes.

I'll also show you a new gravity theory, which has surprising consequences. It treats matter at an extremely small scale as 'spinning light', and that leads to new physics. One unexpected prediction is that if light and matter are sent off at right angles above the Earth (light to travel horizontally, matter to fall vertically) both will hit the ground at exactly the same moment, if they both do. Gravity pulls light, but it's far from obvious why they hit the ground at the same time. Standard theory agrees, without an explanation. The rule was found in 2021. It wasn't found before because no-one was looking for it, but the new theory led right to it (Chapter 33), and has a beautiful explanation.

The important shift *within* general relativity this century is about Minkowski spacetime, on which general relativity is utterly dependent. Spacetime used to be seen as fundamental, as it's about the dimensions. It's now seen as not fundamental but emergent, and emerging from something unknown: that's a major shift, and it has happened very quickly. It's partly because of problems joining general relativity to quantum theory, but string theory also suggested it. It adds an unknown layer beneath what we have, which means all bets are off - physics could go anywhere from here.

But for now I should say something about gravitational waves, and clear up two misconceptions. Gravitational waves have been measured in a new way recently. Because of incomplete or inaccurate coverage, many think the LIGO measurement announced in February 2016 confirmed general relativity, and was the first detection of the waves. Neither is true: the first detection of gravitational waves was in 1978, when they were shown to exist beyond any doubt. The Nobel prize for that was given out in 1993.

And gravitational waves are not exclusive to general relativity. It was Einstein who first suggested them, but they've featured in a range of theories (listed in the reference section, and there's a graph on how the waves were tracked from 1975 to 2005). The detection of 2016 mattered - it used a new method, but it didn't confirm general relativity. It was important for other reasons, to astronomy more than to central areas of theoretical physics.

Thinking about what might replace our present ideas on gravity, Bousso says he thinks the flaw in our current picture "...*hints at profound discoveries and conceptual revolutions still to come. One day, the beautiful explanation that has just transformed our view of the universe will be supplanted by another, even deeper insight. Quantitatively, the new theory must reproduce all the experimental successes of the old one. But qualitatively, it is likely to rest on novel concepts, allowing for hitherto unimaginable questions to be asked and knowledge to be gained.*"

New gravity theories always appear from time to time, but this century the scientific community is taking a lot more notice of them. People are aware that the detailed effects of gravity can appear out of concepts very different from curved space. Two recent theories that have had a lot of attention are Hořava gravity and Verlinde's entropic gravity.

So contrary to the impression that some have had from the non-scientific media, general relativity is widely being questioned. It's happening nowadays in ways that would have been hard to believe even in the late 20th century. What that means is that gravity is, yet again, a force to be wondered about. The apple in the tree is an open question again - and given that it is, it'll best be wondered about, as far as is possible, with an open mind.

JMK

Part 1. Missing jigsaw pieces?

1. An unfinished synthesis

Sitting near some apple trees one day, an apple landed on the ground near Newton. He asked himself - "why do things always fall 'perpendicularly to the ground'?" The apple hit the ground, and immediately afterwards the idea hit him, and came into his head. But people have simplified the story since then, and nowadays the apple hit him on the head.

All it did at the time was to set him on the road to a better understanding, by realising that something we take for granted can be thought about. He soon realised that the force that keeps the moon in orbit, like a circling stone tied to a string, might be the same force that pulls the apple downwards. Then a calculation showed the two forces matched up well ('pretty nearly', as he put it). But the road to a better understanding didn't stop there, and we're still on it.

In a gravitational field, several things happen all at once. Matter is affected, light is affected, what we call time is affected, space is affected - or seems to be. As far as we can tell, *everything* is affected. This is where the good clues about the structure and nature of things are to be found. Rather in the way that Freud studied normal psychology by looking closely at the exceptions to the rule, the place where we might learn a lot about the components of our universe - space, time, light and matter - is where they've been altered from their normal state.

This synthesis of effects has been studied for centuries. We've been steadily filling in the gaps in the jigsaw, finding things out, and doing experiments and measurements that get more and more accurate. But what happens still isn't fully explained. We have a very accurate set of laws for gravity, certainly for solar system type distances and masses. Newton's laws are so good that they provide most of the accuracy, but small adjustments from Einstein's theory make it even better, and help us navigate the solar system.

For decades there were two anomalies in the solar system, but the Pioneer anomaly seems to have been solved, and it wasn't gravity doing it. The flyby anomaly remains, and it's a major puzzle. It has recently reappeared in the orbit of the Juno probe around Jupiter. As the NASA team that studied it for twenty years concluded, it seems to indicate that there's more to be learned about the gravitational fields of rotating objects like planets. If so, that would go against general relativity. There are other anomalies, but we still navigate the solar system wonderfully well.

Our overall understanding has come a long way, but some things are clearer than others, and certain areas are simply not understood. The best indicator of this is that there's a lack of *consensus* on how to explain certain aspects of gravity. These include some very central questions, such as the actual force of attraction (whether a force or a 'pseudoforce'), between two objects. The pull between two objects moving in relation to each other is well explained, but perhaps surprisingly, not between two stationary objects.

Looking at the wider landscape, gravity is like a black sheep among the four forces, in the same sort of way that time may be an ugly duckling among the four main dimensions. Whatever the underlying reasons, it's different from the other forces in important ways. This 'black sheep of the family' aspect of gravity never used to matter. When we were certain space curvature is the cause, it made sense that gravity was hard to classify. In general relativity, it wasn't even a force as the others are, but a pseudoforce, that only *appears* to pull matter. This apparent pull was thought to exist because of where the hidden contours of space happen to go.

So at first it was fine that gravity didn't fit into some description boxes. But it also didn't fit with quantum theory - or anything much else. And as the 20th century went on, it became increasingly clear that our gravity theory also didn't fit into a box labelled 'The standard model of particle physics', which is based on quantum theory. And by the late 20th century, a wider box we have, that had been getting much more clearly defined, came into these questions. We started to realise that general relativity fails to fit with the contents of a box we label 'small-scale physics'.

2. An island chunk of jigsaw

But understandably enough, the experimental results made people very sure general relativity was right. Any theory not fitting with it looked suspicious. Quantum field theory, QFT, was a hugely successful theory: its predictions were confirmed again and again, and it fitted beautifully with other physics. The only reservations some have had, and QFT has even been questioned at times, were because it doesn't fit with general relativity.

So what physics actually does? Almost all of it has been built *from* general relativity. It's an island chunk of jigsaw, but with quite a few pieces in place around there. But it doesn't join onto the mainland. And the mainland, which is largely based on quantum theory, and connects neatly with other physics in many places, has been making very good progress.

And tantalisingly, in the places where gravity actually does join up with the rest of physics, we sometimes don't understand the links. Gravitational mass seems to be the same as inertial mass - we know this is true with an accuracy that keeps on increasing. It has got to at least twelve decimal places now, but we don't know why this should be so. Some people (who may be in denial of a very good clue), hope and expect that the two kinds of mass will turn out to diverge a few decimal places on, making gravity independent, so making parts of our present view work better.

We've been stuck for decades trying make gravity fit with quantum theory, and the rest of what we know. What could be missing from our ideas about gravity? You know what I'm going to say (perhaps from Book I) - a complete conceptual picture. Whether general relativity is right or wrong, either way the accompanying picture is incomplete. To improve on it, and to fit gravity better into our picture, before a mathematical breakthrough, we probably need a conceptual one.

General relativity ultimately fails to join the mainland in an uncompromising way. Calculations for small gravitational interactions, when put together with quantum theory, can give complete nonsense. No-one talked about this until recently, and physicists take it in different ways. Some think general relativity works well within its own domain, as Newton's laws worked in their domain: low speeds and weak gravity. But if general relativity is just an approximation for the large-scale domain, it would need to be replaced.

If it does need replacing, that might explain a long list of issues that are in no man's land at present. The other three forces have clicked neatly into place. There's electromagnetism, and two forces that work at short range only: the strong nuclear force and the weak force. We've found it works beautifully to assume that messenger particles convey them - the photon is the particle for electromagnetism. With three forces, this approach has been neatly tied up: each of the carrier particles has been described mathematically, and found in experiments. So that's the mainland: the standard model.

That approach was looking so good that by the later part of the 20th century, people wanted to do gravity in the same way. So a different picture split off from the curved space picture, with hypothetical carrier particles - gravitons. This is a different way of describing the same picture, treating curvature as a field. It was conceptually weak, because the analogy was not with a concept, but with how we've gone mathematically with other fields, and that in itself lacked analogies. Gravitons turned out to be more or less undetectable, but

two separate chunks of jigsaw about gravity were now building up near each other - from the early and late 20th century.

What they were really about was attempts to build a bridge from both ends. People tried building from the old picture of gravity to the mainland, but couldn't get there. Later on they tried building from the mainland to gravity, with the graviton approach. But the graviton theories had problems, and that bridge didn't join well onto gravity at the gravity end.

The fact that both of these areas exist within mainstream physics underlines the real situation: that neither works well. If either did, the other would go. Still, many believe the path towards a theory of everything includes gravity being inducted into the hall of fame of the standard model. But the truth is, it has been unexpectedly hard to induct it anywhere.

3. Something unknown

Although we have good mathematics that describes gravity, in our attempts to understand it, there still seems to be something unknown at work. A lot of physicists think it's something to do with space and time. There are different approaches, but it's hard to avoid the idea that gravity is closely connected with the structure of space, and with time.

Many, of course, think something more specific: that the structure of space and time is curved into a large-scale extra dimension, as in general relativity. But whether or not that's so, gravity seems related to the landscape of space and time somehow, and it's widely thought that a better understanding of gravity would go with a better understanding of both.

And because a force connected with space and time has failed to link with the rest of our picture, whatever it is we're missing looks important. But until the 21st century, it was more like a vague feeling we were missing something. It's now manifested in the definite decision that many have made, to take spacetime as emergent, not fundamental, and emerging from something as yet unknown. That's a major change, but it has become part of the general view. It's partly because of the way general relativity has failed to fuse with quantum theory, but string theory also suggests it.

Spacetime is *the* place where gravity doesn't connect with the mainland. It's exactly where quantum mechanics and general relativity need to join up, and they don't. Minkowski spacetime is a central part of general relativity, and if it's questioned, the theory is also effectively being questioned. Nima Arkani-Hamed, the well-known American-Canadian physicist, famously said with the opening words of his Perimeter Institute lecture:

'Both these principles [quantum theory and spacetime]*, certainly one them, are going to have to be modified somehow, or changed in a very significant way, in our next description of reality. So I think almost all of us believe that spacetime doesn't really exist, that spacetime is doomed, and has to be replaced by some more primitive building blocks.'*

Spacetime was originally seen as utterly fundamental, because it was about the dimensions. Taking it to be a superficial thing, as we do this century, and emerging from something undefined, is not like saying general relativity is wrong. But it's nearer to it than some people realise.

For now anyway, the bottom line of this is simply: spacetime has been hard to connect with quantum theory. But more or less no-one thinks quantum theory is to blame. Because of this, people have been experimenting with pulling spacetime apart. One of many attempts to do it (Chapter 16) involved separating time off from space at high energies.

So the area with the question marks around it is still space and time. That's the place where the problem is, and also where a solution might be found. And probably for just the same reason, there are missing pieces all the way around this island chunk of jigsaw that we call gravity physics.

It's worth mentioning that experimental results that seem to support general relativity often get people's undivided attention. We're told general relativity has 'passed every test with flying colours', but in fact it has only passed every test mentioned by the people who use that phrase. New observations, which really do seem to go against general relativity, are now starting to be talked about. Some are listed in Chapter 81, p136, 'The strongest evidence against GR, from new technology'.

Another is that an enormous amount of vacuum energy, per cubic meter of space, was predicted by quantum field theory. QFT has been right about all its predictions so far. Later, this amount of energy was effectively measured and confirmed by experiment, via the Casimir effect. So the idea that all this energy exists is very strongly supported. Now according to general relativity, energy *always* creates gravity. It's an important part of the theory, so all that energy should create a lot of extra gravity. But it's not there. If this gravity existed, the universe would be very different. For one thing, the solar system would fall apart.

So it's clear that although all this extra gravity from the vacuum energy must either exist or not exist, either way, we're not going to observe it. If general relativity was right, and all this gravity existed, we wouldn't be here. So the fact that the solar system is in good shape, and we are here, shows general relativity is incorrect. But looking on the bright side, you might say a gravity theory is a cheerfully small price to pay in the circumstances.

4. A funny thing about interferometers

There are a few misconceptions about the recent detections of gravitational waves. These include the idea that the measurements have only one possible explanation, and that they confirm general relativity. For one thing, there are many theories that predict gravitational waves (papers are in the reference section). So there's a range of possibilities from physical theory, but there's also an ambiguity in the measurements themselves.

Interferometers like LIGO are the descendants of the first interferometer, as in the Michelson-Morely experiment. And from its invention in 1880 to the present day, the technique - of sending two light beams off at right angles to mirrors, to return and be recombined - has had an ambiguity about it.

There have always been two possibilities that are hard to tell apart. Firstly, interferometers could be measuring variations to the speed of light, as it was hoped they would in the Michelson-Morely experiment. But secondly, they could be measuring differences to the length of the interferometer arm, as it was hoped they would in the LIGO measurements.

The well-known early null result proved that the aether didn't exist, and was unambiguous - Michelson and Morely measured neither. But the present day experiment has found something trillions of times smaller than what failed to show up 130 years ago, and it absolutely could be either. The ambiguity exists because the setup effectively times the light on its journey. If that time period shifts, either the light's velocity has briefly changed, or the distance it covers. So in $v = d/t$, if t shifts, either d or v could be varying.

And the idea that interferometers measure *speeds* for light is far from being an odd idea. It's the opposite. Nowadays interferometers are routinely used to measure the refractive index of gasses - we find out how fast light travels through a gas by building something just like LIGO, but smaller. We also use miniature LIGOs to measure the temperature and pressure of gasses, and assume that the light's travel speed varies. In fact, we know it does, beyond any doubt. But at LIGO itself, which tests for variations at a smaller scale, we assume an entirely different principle is at work.

In the theory this book is about, and in general relativity, gravitational waves are ripples in a medium. In both theories, it's a medium closely connected with space itself (unlike the aether, which was thought to be like matter). But there are differences: if the medium ripples, in one theory distances ripple, but in the other the medium's *density* does. This means that travel speeds also do. You get two alternative equivalent pictures, and they're so similar mathematically that the measurements look the same.

Only one of these two principles is tried and tested. It's well known that if a medium like air or water varies its density, light travelling through it will be slowed slightly more or less than usual. The density can 'ripple': this happens all the time on Earth.

So technically, it's less of an assumption to say that speeds are what varies, because we know this happens. But whichever assumption is more of one, it could be either. Something is rippling, that's clear enough, and we know the mathematics it goes by. There's more about the alternative version later on, and Part 3, 'General relativity's secret twin', shows a reason why such a close similarity between the two theories could exist.

I needed to go straight to gravitational waves, but now I should start at the beginning, and tell the story. The lateral picture that emerges leads to new possible answers for some unanswered questions about gravity. And it leads to new possible answers for some answered questions as well.

But leaving the wider questions and zooming in, as we must, whatever else is going on, gravity is about matter. For some reason, all matter attracts other matter. This affects every different kind of matter in exactly the same way, whatever state it's in, and whatever it happens to be doing. So gravity looks more fundamental than whatever state it's in, and whatever it happens to be doing.

The more mass that's clumped together in one place, the stronger this force that pulls things towards it will be. And the more mass, the stronger all the other effects to do with gravity are as well, including a slowing of time itself. There's a closely interlocking set of effects, and so far no-one has managed to explain them all fully, and how they fit together.

Part 2. Unanswered questions

5. What we don't know

Newton found the broad, basic laws that describe gravity, with a set of clear, simple equations. Each had just a few terms. They worked so well that they seemed to be basic elements of the universe, but he didn't know what the underlying cause was, or what he was describing. Then, in the twenty years that followed, he had some wonderfully good ideas about what was really going on. But important bits of information were missing, and it was simply too early to get a complete view of things.

When Einstein found the idea of curved space a few centuries later, it led to an even more accurate theory, which made fine adjustments onto Newton's equations. It also did a lot of other things, and it did them enormously well. It described gravity in detail, and was right about the details again and again.

General relativity deals with gravity very well, but its success is in describing gravity's effects. When it tries to describe its causes, it does far less well, and it has very incomplete explanations in some areas. These include the force of gravity itself.

By intuition, without looking closely, curved space might seem to cause this pulling of one mass towards another. It seems that we can think of an object released near a mass as in the trampoline analogy: a ping pong ball released near a bowling ball will roll across the trampoline towards it. It might swing around the larger mass, and follow the curve of the indentation in the sheet, travelling around it.

But it's well known that this picture can be misleading, and doesn't do what it appears to do. A ping pong ball released on the trampoline will only move if gravity is already at work underneath the trampoline.

To be on the safe side, best to put the trampoline out in space, far from any mass. Objects released on it then wouldn't roll anywhere. The ping pong ball, when released, would simply sit still. So the picture doesn't explain gravity, as it depends on pre-existing gravity to work at all. It just helps us to visualise things. That in itself isn't a problem, it's only an analogy. But there are gaps in our understanding of the real world as well.

6. Why does an object start falling when released?

One of the gaps is about why an object starts to fall when released. The idea that the presence of a mass curves the space around it does a whole range of interesting things. But it doesn't explain why a stationary object released in the field suddenly starts moving, and sets off towards the mass.

But on the other hand, once in motion, curved space describes precisely how an object will move. Moving objects follow paths through the indentation in space, which are like the contours of the trampoline. General relativity traces out these geodesic paths that curve around the mass in four dimensions, and the mathematics works beautifully once the object gets started.

On Earth, with other forces at work to confuse matters, motion tends to be a temporary state anyway. But in space, objects go on moving. So the paths followed by moving objects are more fundamental and important than they might seem. Curved space does very well with objects in motion, but it can't explain why an object released at some height in the field will suddenly start moving downwards. Space can be curved, flat or rabbit-shaped, but nothing about its shape alone provides a reason for an object to set off suddenly in a particular direction.

This problem with how the motion gets started - the 'starting problem' - has had different reactions. There's no consensus on how it should be explained, and no agreement on any standard answer. I've known people to say it arises because of what's known as 'geodesic deviation', but that's actually about objects that are already in motion. Others say the apple starts to fall because although it's not moving through space, it's moving through time, so there's already motion happening.

But many don't accept that idea, because of a rock solid, rigorous proof that has effectively been taken into relativity, discovered fifty years ago. It proved motion through time doesn't exist, certainly not in the way that's required for that idea, if spacetime *does* exist. And general relativity depends entirely on spacetime. Some physicists who try to uphold general relativity seem to have motion along the time axis existing when it suits them, such as in order to solve the 'starting problem', but not when it doesn't suit them.

I've known others to say that gravity is the difference between two reference systems - but that doesn't make objects move. Anyway, however you take it, the question is controversial. Some physics sites are covered in complicated discussions, and sometimes arguments about it. And as always, the breadth of the spread of answers shows the extent of the problem.

Some top relativists - people seen as having some authority - have said that the real mystery is not why an object falls, but why matter can stop it from falling (such as a hand holding an object before it's released). Or why hitting the ground stops it from following its natural geodesic path, as if all matter should really be in motion all the time. This would helpfully place all matter in an arena where we're able to describe what it does.

But very few try to argue that our understanding of gravity is complete, and the general view within the scientific community is that gravity is not fully explained. This problem of how the motion starts is one of the reasons, but it's linked to an even wider one: what causes the actual force of attraction between objects. In particular, two stationary objects.

We intuitively believe that when the apple is still in the tree, even though it doesn't move, the force of gravity is pulling at it. Because general relativity can't explain that easily, and there's no consensus, there's a need to see if it's true. After all, in physics intuition is often wrong. So how can we prove that there's a force pulling at a stationary object? Well, the ocean stays in place, but the horizon curves. Different parts of the ocean are being pulled towards the centre of the earth, and kept in position by a force.

In general relativity, this pull that keeps the ocean in place is still a mystery, although orbits, and how light moves in a gravitational field, on the face of it are extremely well understood. But describing things accurately is one thing, understanding them is another. With orbiting objects and curved space, the objects are thought to move *as if* with a pull towards the centre, rather than actually with a pull at work. The paths that matter takes effectively simulate a pull towards the centre without the need for a real force, because they're thought to be the natural paths through four dimensions.

Because we can trace the paths, it seems we've also explained the apparent pull, and shown it not to be real. But the problem is, in other situations, such as where a released object starts to fall - or before being released, when it's held in place - there's still a pull towards the centre. How that pull works has been very much harder to explain. Some think it has been explained, others think it hasn't. But the prevailing view is that the force of gravity is so far only partly explained.

And in general, an explanation that works well in one area but not in another is a questionable one. A good explanation should provide a range of answers, that dovetail together neatly in many places. But even after a hundred years, curved space has a shortage of answers in some key areas. This might mean that general relativity is wrong, because in physics, two different pictures can produce similar mathematics. So the correct picture might be different.

7. Early false ideas

Before Galileo's time, there was widespread acceptance of Aristotle's view of the physical world. Aristotle was a bold philosopher, and he worked on a very wide range of subjects, but in physics he had a bit of a knack for being wrong about things. In some areas it rather became his trademark. He was also very influential, and people didn't check his ideas, they just took them to be correct. This held back progress for a long time.

Two thousand years later, around the 17th century, the Renaissance brought a new excitement about observation. This was the beginnings of the modern scientific method, and when people finally started to look for themselves, they found out just how wrong Aristotle had been about everything. Galileo wrote a book that rejected Aristotle's views on motion. Contrary to the old idea that heavier objects fall faster, he found that all matter falls in the same way. And Newton found that colour wasn't a mixture of black and white, white light was a mixture of colours. And he found that when matter is moving in a straight line, it keeps moving unless a force is applied to it. This was also the opposite of Aristotle's wrong view.

And it was Aristotle's idea that nature abhors a vacuum. We now know from observing the universe at large that more than 10^{82} cubic meters of vacuum have arisen in nature, so although he meant something else, he produced a candidate for the most false statement ever made. But the overcritical view of Aristotle is misguided (even Shakespeare laughs at him), and the bad press he's had since the 17th century is unfair. And though an error about his work was perpetuated for 1000 years, corrected, then - after another 1400 years - mistakenly mocked on the internet, it seems that he thought mayflies have six legs, not four, if you include the ones not used for walking.

Aristotle probably spread himself too thin, he studied just about everything open to study at the time. But it was the Greeks who came up with the idea of thinking about things at all - that is, in a particular, ordered way. We take that for granted now, but it needed inventing at the time. Aristotle was one of the first people to ask the questions, and we can't blame him for not going straight to the right answers. He was important in setting humankind on the road to finding the answers, but because people assumed that this first set of answers was correct, there was a bit of a delay on the road.

8. A modern false idea

That delay happened because science was invented before the later system of checking everything by experiment, which came in the Renaissance. Since then we've confirmed what Galileo found out with more and more accurate experiments. All objects, whatever they're made of, accelerate in the same way when gravity pulls them.

It's said that matter's consistent response in a gravitational field is evidence for curved space. This point is quite commonly made, and is found in many places on the internet. Perhaps when matter goes near a mass like a planet, all different kinds of matter behave in the same way because the background space takes a different shape near the planet. At first glance this can look like a very reasonable argument, but it's not what it seems, and it fails.

The first point to make is, matter responds to forces in a similar sort of way elsewhere, in areas outside gravity physics. And it turns out, looking closely, that it's actually responding in *exactly* the same way. In deep space, far from any gravity fields (I'll use the informal term 'gravity field' from here), matter accelerates when a force is applied, going by Newton's key equation, $F = ma$. This happens in a general way. It happens with wood, metal, you name it, it makes no difference. Matter of any kind accelerates as in $F = ma$, that's just how matter responds to a force.

On the face of it though, this looks different from the uniform acceleration due to gravity. With $F = ma$, the acceleration isn't uniform, it varies. Make the mass three times larger, but keep applying the same force. The resulting acceleration is now three times smaller. With its inertia tripled, the mass is now harder to push, and the resulting acceleration changes. So this looks like something different.

But it's exactly the same. That's because you don't keep the same force, as you do in the above example. With gravity, make the mass that gravity works on three times larger, and you inevitably make the force at work on it three times larger as well. The size of the mass affects the force. We know this, and we know about the basic cancellation it causes, leaving the acceleration the same. Physicists can choose whether or not to mention it.

But how can this be exactly the same as $F = ma$, surely gravity is different? If you look at Newton's simple gravity equations, they're in a precise $F = ma$ relationship. The formula for the force F, equals the mass the force works on, m, multiplied by the formula for its acceleration, a. The end result is $F = ma$. F (or GMm/r^2) = m by a (or GM/r^2). Newton probably reached them that way, though the order he did things in makes no difference.

The point is, if a force is applied to a mass, it causes an acceleration, affecting all matter via $F = ma$. This is true even if the force is gravity. Gravity varies slightly at different heights in the field, but at any given point, $F = ma$ can be seen at work. And Newton's gravity equations are far from outdated, some of them are functional parts of general relativity.

The uniform response of matter to forces is a fundamental thread that runs through many areas of physics. It happens in interactions, via energy, inertia, and so on. It happens with or without gravity involved. So when it happens with gravity, you simply can't say that it's *evidence* for curved space. If it was explained by curved space in the context of gravity, we'd still need another explanation for it - *everywhere else*. And because it looks very fundamental, what's needed is a wider explanation.

9. A common root

That particular argument for curved space also fails because it's out of date. It used to fit Einstein's view, who reached the idea that space is curved partly via the fact that all matter falls in the same way. But since then, a long list of reasons has appeared to think that all matter has some *common root*.

It's not just mass and inertia, though the uniform behaviour of matter via its mass is important. There's also the conservation of energy, and other recent clues. A new cluster of clues arrived in the '80s, with string theory, that also suggest this very strongly. Some of the clues were there early on, but back then they looked like they might eventually get explained in some other way. Meanwhile, the list of reasons to think that all matter has the same nature at some level, and a common origin, kept lengthening.

This went on until it was unavoidable, three decades after Einstein's lifetime, with the Planck scale physics of the late 20th century. What it means is that nowadays, the fact that wood and metal fall in exactly the same way doesn't *necessarily* result from the geometry of space, as it seemed to a century ago. The idea that it necessarily must be about curvature is out of date. It might, but there are now other possibilities.

To sum up, there are two general concepts, either of which might explain matter's uniform response to gravity. The two headings are: 'curved space' and 'common root'. But all matter having a common root would explain the uniform response more widely than curved space, which only explains it for gravity. The common root idea might explain it for $F = ma$ everywhere. That would cover a lot more, and it could include gravity, given that the equations of gravity, and its behaviour, are exactly as in $F = ma$.

Part 3. General relativity's secret twin

10. A built-in possibility

With what's called the equivalence principle, we take on trust a simple idea. It's that the effect of gravity is indistinguishable from an acceleration. There's more than that of course, in the ideas surrounding the equivalence principle, which are at times subtle and complex. But there's no need to go into them very far, to make a point about an alternative possibility.

Einstein saw that an acceleration in flat spacetime is like the effect of gravity in curved spacetime. One is matter responding to a real force, the other is matter following the contours of spacetime. the result behaves like a force, but is a 'pseudoforce'. Over a small enough region curved spacetime is more or less flat, so it's impossible to tell the difference.

A central part of that is the idea that gravity is equivalent to an acceleration. But what many people don't realise is that if by any chance, instead of being equivalent to an acceleration, gravity actually *is* an acceleration - of the kind caused by a real force - you'd get nearly the same mathematics, but general relativity would be wrong.

There's clearly some link between gravity and acceleration. There are issues about reference frames I won't go into, but general relativity assumes that the link is indirect, and a very close similarity, as it may be. But the fact that this similarity is part of a *principle* has a habit of pre-empting any discussion about it as a clue, or as an indicator. I'm not saying that this was intentional, because it wasn't. But it has had that effect.

It very often makes people take all the clues that suggest gravity actually *is* an acceleration, and ignore them, or bin them. But people didn't ignore them in the early 20th century: physicists like Einstein knew that gravity behaved as if there was a real, direct force at work. This would explain a list of key clues, including the $F = ma$ in Newton's equations, and the equality of inertial and gravitational mass, which is still seen as a mystery nowadays. If a real force was at work, of course the two kinds of mass would be the same. Both would simply be the 'm' in the $F = ma$ of the force of gravity.

But people couldn't find a mechanism for a direct force, either because there wasn't one to find, or because it was too early to find it. If it existed at the

Planck scale, for instance, they wouldn't have had enough of a picture to get there back then, as the Planck scale was only a limited set of hints and vague ideas at the time.

So they found another way, and decided that there was a 'fictitious force', or pseudoforce at work. These exist elsewhere in physics, and have aspects that link to acceleration, in ways that I won't go into here. Suffice to say that the view that emerged was a perfectly valid one of course, and self-consistent. It also fitted, rather loosely in places, with gravity as we observe it.

11. Not a coincidence

So there were two alternative ideas behind the setup that Einstein worked with, which later on was so successful in describing gravity. General relativity contains this double picture, and a potential ambiguity, because of the early starting point. As in Einstein's unmistakable style, the starting point involved just a few simple principles.

And one of the absolutely central principles was actually *about* an ambiguity - the equivalence principle. It was about the unexpected equivalence of two pictures. The second possible picture arises from that original schism within the conceptual foundations of the theory. It allows the mathematics to be equivalent in both, because although Einstein chose to cast general relativity in the form of space curvature, loosely speaking, he could choose how the curvature was 'set', and he set it to match certain ideas. And he used the other picture as an early reference point.

I'm not saying that in all areas, the two things were in fact exactly the same all the time. I'm just saying general relativity's starting point drew on certain areas where that idea applies - and as a result, the mathematical description can be equivalent to, or close to, another conceptual basis.

Einstein talked about this equivalence at least as early as 1907, and it led to general relativity a decade later. In between special and general relativity, his clear thinking led him to all the links, which started out from the fact that all objects fall in the same way. He wrote: *'It is only when there is numerical equality between the inertial and gravitational mass that the acceleration is independent of the nature of the body.'* And having decided to build one picture out of the other, he wrote: *'We [...] assume the complete physical equivalence of a gravitational field and a corresponding acceleration'.*

These steps are seen as insights into the nature of gravity, and they certainly were. They made sense of the clues that were available at the time. But they didn't necessarily lead to space curvature. They led, initially anyway, to the idea that a real force was at work, behaving as other forces do, and going by the $F = ma$ visible in Newton's gravity equations.

But the problem was, no-one could think of what that force could be. There wasn't a force around that could do it. And it wasn't a straightforward force - there were still issues about reference frames, and other things. Einstein took another decade to complete a theory that mimicked what those clues seemed to be saying, and which also fitted with special relativity.

So general relativity's starting point had two interpretations, and the ideas are like twins separated at birth. One was that gravity is like an acceleration, the other was that gravity is one. One twin went into general relativity, while the other waited quietly in the wings, perhaps until such a time as physics might develop far enough to find the mechanism.

What originally gave birth to these twins? Experiment, starting from Galileo's discovery that all matter falls in the same way. That led, as Einstein realised, to the two kinds of mass being the same, for whatever reason. And there the road forked, and split into real force or pseudoforce - the two alternatives. The fact that the other twin potentially exists means a different theory might mimic general relativity in the experimental results.

Without this duality in the concepts behind general relativity, the idea that a theory exists that's similar enough to *replace* Einstein's theory is possible, but it looks far-fetched. The experiments make it look like general relativity has to be right, and although we now see it as an approximation, it seems unlikely that anything else could do so well. But it's far easier to believe that such a theory exists, and that general relativity might be false, if one knows that it doesn't necessarily even require a coincidence.

So here in the 21st century, now that we're finding major reason to question general relativity, these clues about an alternative become interesting and relevant. Curvature is failing to work with the rest of physics these days. But the accompanying mathematics came out of two pictures rather than one - and the other one might still be right.

12. How do we choose between these pictures?

Before describing a twin theory, I should distinguish between it and another group of theories, that were on the fringes of 20th century physics.

There was quite a long gap between relativity's acceptance, and its accurate confirmation via advanced technology. That gap was about fifty years. During that time, some rebel physicists thought it was wrong, and they questioned the experimental results. Their theories assumed that there would be major measurable differences, but they were proved wrong after new technology arrived in the 1960s. Some clung on, but possibilities of that kind were dead. The theory in this book isn't part of that pre-string theory approach, at which point a lot of things changed - from the '80s on. Its predictions are essentially the same as those of general relativity, so it's a different animal, and it comes out of a closely equivalent picture.

The points made over the last two chapters are very general, and certainly include a number of oversimplications. The foundations of general relativity are subtle, as are the surrounding issues. I haven't gone into them in detail, because there's no need to, in order to make a particular point that applies regardless of the details.

If you say that 'another conceptual basis might exist', the number of possible theories you're talking about is enormous. Of course, they won't all fit the bill, and whether any of them do is another question. When one starts being specific and setting out a theory, as I will later on, certain things need to be dealt with. But for now, saying the possibility exists is partly just a reminder about an interesting aspect of the equivalence principle.

So at the starting point, we have these two similar pictures. There might be a theory for each, so the question is how we choose between them. One is far more complete than the other, but several things might help with trying to assess and compare them. And it could be a vital question: although the two start from a place of close similarity, they may lead to different paradigms in physics, and the end results could be worlds apart.

Firstly, we have a neat set of three interlocking concepts. (1) All objects fall in the same way, which leads to (2) inertial mass and gravitational mass are the same. And that leads to (3) gravity is either A. an acceleration, or B. very like one. So which is it, A or B?

If any loose ends are left sticking out, when we tidy up these few interlocking concepts, that would look suspicious. And the fact is, in general relativity, it's still a mystery to this day that inertial mass is exactly the same number as gravitational mass. We don't know why this should be, but experiment keeps showing it to be true - with a steadily increasing accuracy, over a long period of time. So far it has got to at least twelve decimal places.

But the other twin ties up all these concepts neatly, with no loose ends. The two kinds of mass are the same because a real force is at work, and being a real force, it affects all the different kinds of matter via its inertial mass, with the usual relationship between force, mass and acceleration.

What else is there, to help choose between them? General relativity also has trouble describing the known behaviour of gravity, when it tries to paint it as the result of matter following the contours of space. It doesn't explain the force we know is at work when the apple is still in the tree, and it has trouble with what happens when the stem breaks. And relativists disagree amongst themselves *a lot*. So one of these sets of concepts has conceptual problems. But the other twin says a real force is at work, and that can be the force we know is pulling between the Earth and the apple before it falls.

So the very same idea might solve several separate problems. That's always a good indicator, but it has to be admitted that the other twin is enormously incomplete. It always was: it was just the idea that there's an undiscovered real force. But it would make the whole jigsaw fit together a lot better, and perhaps that much can be seen already.

And talking of jigsaws, that brings me to the point that general relativity is like an island - it doesn't connect well to the rest of what we know. The other twin has a long way to go, and has only made a few connections, for instance with other forces. But it still might give us some pointers on where to look, and perhaps about where one might find the beautiful future theory that Raphael Bousso described in the quote in the introduction.

What clues, for that? Not many. Best to search in areas of physics that have been found more recently, such as Planck scale physics. If it had been around for longer, there might have been no need for a 'pseudoforce' at all. And whatever the mechanism is, it works in a way we know about already, in one place at least. It affects things locally via $F = ma$, and as usual in that setup, all matter responds to it in the same way.

Anything else? Well, general relativity itself should be absolutely packed full of clues, given that we're trying to find out about its missing twin. After all, if there's another set of ideas that's the true picture, general relativity was still shaped by it, which is why it keeps getting things right.

Part 4. The plot thickens

13. Newton's gravity constant

There are other clues that can be looked into, to help with choosing between these two pictures. The alternative picture becomes increasingly noticeable on the way, but only if one is open to seeing a hole in the jigsaw, and open to the idea that there may be a force at work that is so far unknown, but which nevertheless has some recognisable characteristics.

Physicists are sometimes not very good at leaving holes in the jigsaw, and so temporarily sacrificing the comfortable feeling of knowing what on Earth is going on. But leaving holes helps with progress more than anything else, and filling them (before we're ready to) blocks progress. The physicists who make breakthroughs have always been aware of the unknowns, and have left room in their picture: new ideas need somewhere to land. Perhaps the greatest single bias in physics is to believe the standard view, and ignore the places where it doesn't fit, as it's the best theory we have at present. That doesn't make it true. And to put that in perspective: so far every single established, central theory, without exception, has turned out to be wrong.

To see the gap more clearly, we need to take a look at the gravity constant. Newton's constant is a number used in almost any gravity calculation. More or less everyone working on gravity since Newton's time has used it.

Gravitational mass is one of several kinds of mass, and it subdivides into two further kinds. Active gravitational mass is about when matter causes gravity, passive gravitational mass is about when it responds to gravity. They're often simply known as active mass and passive mass, or 'the mass' and 'the small mass'. Calculations using Newton's theory tend to have a central active mass, like a planet, and a smaller passive mass, like an orbiting object. Newton set his constant so the same value for an object's mass can always be used. But the fact is, for whatever reason, we often divide the value of a mass by a very large number, about fifteen billion. Or rather, we multiply it by a small one. That gives us the right answer, but it doesn't explain why.

So in Newton's force equation $F = GMm/r^2$, what is the constant G actually doing? It comes once, but there are two masses. So if it's being multiplied by a mass to make an adjustment, it's only being multiplied by one of them. Elsewhere G often appears with just one mass. It's always an active one. The active mass M is never seen without its constant companion, G.

Newton's law is stated as if the masses are on an equal footing - *'the force is proportional to the product of the masses, and inversely proportional to the square of the distance between them'*. This is true of course, but there's still an active mass and a passive one. The two are interchangeable: the force pulling between them is the same either way. The equation gives the same number because of a symmetry in the situation: $GMm/r^2 = GmM/r^2$. But G can still apply to active mass only, in all versions of the situation.

On a discussion site, someone set out some clear mathematics showing that $m_1^{passive}/m_1^{active} = m_2^{passive}/m_2^{active}$. As objects m_1 and m_2 can be different sizes, he pointed out that *the ratio between active and passive mass has to be a universal constant*. If he thought this constant might be G, he didn't dare say it. Supporters of general relativity came in and said that passive mass doesn't exist in their way of doing it. In fact plenty of relativists use it, it can of course be used, and the point still stands.

On that site they assumed the two masses have to be moving in relation to each other, but they don't. A calculation can be done with a dumbell-shaped double mass, with a connecting rod. This is like the apple before it falls: the connecting rod is the tree. But if relative motion between the two masses is prevented, general relativity has trouble explaining gravity.

This point about a universal constant has been confirmed by experiment. The active-passive mass ratio, whatever it actually is, was found to be the same in two different substances. And it becomes clear that this ratio is G. There's a long list of reasons to think the constant is for active mass only. But some relativists who happily talk about active and passive mass, talk as if G might be spread between the two, and seem to dislike the idea that it's for active mass only. It's more or less unavoidable, but talking to physicists, sometimes there's what I'd say is a reluctance to acknowledge it. This is not a conspiracy (and nor is the title of this section!).

Some of the above is an oversimplification. G comes into a lot of equations, and one can't always say what it's doing. But in general relativity itself, G is a coupling constant between matter and curvature. Clifford Will is just about the world's top relativist, and he tells it like it is: *'in general relativity, G plays a role as a constant of proportionality that determines the amount of space-time curvature produced by a given density of matter.'*

If this is so, it makes G a constant for active mass, and supports the idea that only active mass uses it. In general relativity, active mass is matter's ability to curve space: its value needs to be made very small via the constant, because matter only curves space a little. So exactly as Will says, G is a proportionality constant. And it fits the landscape well if it's the ratio between two separate effects of mass - one large, the other small.

Now it doesn't weaken general relativity at all to say that active mass uses the constant. But it strengthens the alternative view, which is sometimes like the elephant in the room, *to say that passive mass doesn't.*

That's because if passive mass doesn't use the constant, then passive mass does indeed behave exactly as ordinary inertial mass behaves, without using G. After all, elsewhere in $F = ma$, there's no G. So the $F = ma$ configuration in the equations would look even more significant. But if G gets spread across both masses, as some say could still be the case, it blurs the issue.

How does one probe this question? Well, the small mass, m, is cancelled in some situations. In the formula for acceleration, the ambiguous 'GMm' goes away, and you're left with just one mass: $a = GM/r^2$. It's an active mass, and it's using the constant, exactly as if only active mass was using it in the other more ambiguous equation.

But it gets better: light has no mass, and doesn't use the constant. It still gets pulled off its path by an active mass when it passes one, even in Newton's theory. That's gravity at work. The deflection angle can be calculated via just the acceleration due to gravity, $a = GM/r^2$. *So there you have an active mass using the constant, pulling something passive, that isn't using the constant, towards it - with the pull of gravity.*

This sets off a train of thought that is like the elephant in the room in physics. The train of thought goes like this. The equality between gravitational mass and inertial mass doesn't necessarily have to be the major mystery that it is in general relativity. Instead, passive mass might genuinely go by $F = ma$, and be the same as inertial mass because it *is* inertial mass. And that would mean there's a real force at work. And that would rule out a pseudoforce, arising from matter following the contours of curved space. And that would rule out general relativity, explaining why it has been impossible to combine it with quantum theory.

14. What kind of explanation is likely?

If one looks at the uniform response of matter to forces, it starts to suggest a few things. It happens consistently, with matter in any state, in a huge range of situations, and it happens via matter's mass. It suggests there's a place somewhere, if we could peer down to a deep enough level, where all matter somehow arises from the same root. This chimes with string theory, where all the different particles are the same basic object, but vibrating in different ways. So that's one way matter might be linked up.

The conservation of energy shows it clearly, letting us know that something we don't understand at all links up the different kinds of energy. They differ widely, yet they can act together, keeping an object's *total* energy the same, even if within that total, quantities for the different kinds vary and shift, as energy gets transformed from one kind to another.

But if the total energy changes, something revealing happens. The object's inertia does too, in exact proportion. So this total starts to look like a real quantity, for something physical, not just a mathematical total. But if so, the different kinds of energy must be linked in some unknown way. Where might this link be found? Energy is a mystery, but a long list of clues point at the Planck scale.

And secondly, the uniform response of matter suggests gravity is less about the background space, and more about *matter itself*. Whatever gravity does to matter, it works on it at this root level at which all matter is part of one thing. This level is hinted at by matter's uniform response, including in areas outside gravity. And it seems the explanation needs be one in which matter's nature is important. Rather than being dragged arbitrarily into the dips and valleys of curved space, matter needs to be an integral part of the process, and get worked on at this root level somehow.

One reason to think this is that with gravity, the small mass m affects the size of the force that pulls. An apple five times larger gets pulled by a force five times larger. So it seems that the small mass, although we call it the passive mass, is somehow part of the process, rather than being pulled into the dips and valleys in an arbitrary way. We get the impression that matter isn't very involved, because of the uniform acceleration. But that's misleading, as I've already shown. And the way the passive mass adjusts *the actual force* shows that something else is going on.

15. A glimpse of 22nd century physics

Ask about something we don't know, and physicists will often say 'we don't know that'. What they say far less often is 'we don't that, here are the clues'.

So we can imagine this level at which you get uniformity across all different kinds of matter. What clues do we have about that place? Not many. But this uniformity seems to be connected to matter's mass. Mass is closely linked to inertia, so it's worth looking at inertia for any clues.

Inertia is an object's resistance to changes in its motion, or its lack of motion.

Newton called that an 'innate' property of objects, so he saw it as something internal. Objects resist forces in proportion to their mass. So on the face of it, it looks like something internal. They also resist forces in proportion to their internal energy, which is a stronger reason to think it.

But for a century, the standard view on inertia has not said much about the internal side. Mach's principle, the best explanation we have at present, says that inertia is a very long-range effect, caused by the gravity of all the distant galaxies combined, telling matter how to respond to forces. This approach has been questioned for most of the 20th century, and by some very different kinds of physicists, both rebels and hardliners. It has had various problems, and some say it's inconsistent with general relativity.

Mach's principle has always been like an extra bit tacked onto the main body of physics, which doesn't quite fit, and raises difficult questions. Einstein and Mach respected each other's work early on, but they both because less keen later, when some inconsistencies surfaced. Half a century later, a conference on Mach's principle in 1993 showed there are quite a few different versions of it. Some of them were inconsistent with each other, whether or not they were consistent with anything else.

Mach's principle is still officially the standard view, but since the 1990s there have been attempts to replace it with an explanation that paints inertia as a local, short-range effect, which arises from the quantum vacuum. These new ideas see inertia as caused by the fleeting particles that appear and quickly disappear in the vacuum of space. The idea is that they slow anything down, rather like rain on a car windscreen. This affects moving objects, particularly when they try to change direction. These ideas work in a limited way, but they don't give anything approaching a full explanation. They haven't done well enough to replace Mach's principle, but they *have* done well enough to show that in the future, inertia may turn out to be a short-range effect. At present no satisfactory explanation exists.

But the point is, in our general view of things, inertia is shifting from a long-range effect towards a short-range one. And it's also happening with mass. finding the Higgs boson in 2012 meant that the Higgs field certainly exists in some way, and takes mass (for the elementary particles) straight to being a small-scale, local effect.

In fact, mass has overtaken inertia. And although the two things are so close that they may well be different aspects of the same phenomenon, at present the standard view sees them very differently. Right now, we take inertia to be a long-range effect, involving the distant galaxies, but we take mass to be an effect as short-range as the Higgs field.

But the point I'm making is that this is part of a general drift towards small-scale physics. Inertia will probably catch up. That may give a hint about what 22nd century physics will look like. We're heading towards the place where all matter is the same - that's where this road goes. And gravity, which is closely linked with mass and inertia, may need to go that way as well.

16. Other gravity theories

So there you have a set of clues about what kind of explanation for gravity is likely, and a bit about what the background framework, on which it should fit neatly, looks like. Other gravity theories are not providing what I've said is needed, which is an explanation for the link between all the different kinds of matter. The explanation we need should hook in with other clues we have, for instance about mass, inertia, energy, and so on.

I won't go far into talking about other theories, but two of them have had a fair amount of attention recently, and that supports some earlier points I've made. The interest in Hořava gravity in 2009 led to a meeting of some very good physicists at the Perimeter Institute in Canada at the time. It showed that people are questioning spacetime nowadays, because the theory takes spacetime apart at high energies. This is to some extent part of the general attempts to alter spacetime into an emergent phenomenon.

One of the ideas behind Hořava gravity is that space and time are not bound up together, as was thought, until you get to lower energies. Further up the scale you get something like Newtonian time instead. Like many attempts to bridge the gap between the two main theories, this tries to make quantum mechanics fundamental, and relativity emergent. That's the only way round that looks possible, to make the two compatible. But like many other current approaches, Hořava gravity is about making a comparatively minor tweak to the present picture, to try to put things right. However, we may instead need to replace the present picture we have with a truly new picture, rather than looking for a small adjustment to fix up the old picture, which is what very, very many people are doing now.

Then there's Verlinde's entropic gravity, which tries to provide an underlying framework. It uses what's known as the holographic principle, which some think is unnecessarily complicated (Neil Turok's call for simplicity in physics is in the next chapter). It's a description of gravity in terms of information. The field of information has shown that it can substitute for existing mathematics in some areas, including quantum theory. The same processes are described via parallel mathematics, and Verlinde claims to have recovered Newtonian gravity out of information considerations.

To me, this is likely to be a valid parallel way of describing it, but that doesn't mean that it provides any real *cause*. In the past, we've often found parallel mathematical systems that lead to the same equations. Sometimes you even get a whole parallel framework, like that of information.

There's a very close analogy for this. In Newton's time, people were learning about gravity, and oddly like nowadays, they were getting to the equations, but they didn't have a cause that really worked. Leibnitz, who was Newton's rival, found that he could get to similar equations, by starting from a concept that created a parallel mathematical system: energy. It was possible to trace what happens with gravity via the energy, just as it's possible to trace what happens with gravity via the information. Neither provides a cause, but both provide a mathematical account of what's going on. So until things are better understood, either can seem like a cause.

Leibnitz came up with a gravity theory based entirely on energy. He thought energy is the cause of gravity, and called it the vis viva theory, which means 'force of life'. One equation from it, the vis viva equation, is still used a lot by NASA. It's an orbital speed equation for all different types of orbit, and it was arrived at via the orbital energy. It provides a good account of what happens, but we now think Leibnitz was wrong to see energy as the cause of gravity. It's a good way to keep tabs on what's happening, because everything that happens is *registered* by the energy of the system. But nowadays we don't think it's the cause.

In my view information is a modern parallel, as everything that happens is registered in the information content of a system. This is an important point, because the information approach to quantum mechanics is seen by some as a way forward, although we have no interpretation. It's a highly complicated approach, in some areas. In my view information, just as energy did before it, will provide a mathematical account, but not an explanation.

17. Search for simplicity

The universe has a habit of turning out to be bafflingly, unexpectedly simple. The title of this chapter was the phrase of a well known, respected Professor of physics at MIT, Victor Weisskopf. He spoke out many times in favour of a conceptual approach to physics. His phrase 'search for simplicity' was about looking for the underlying concepts, and what they're often like.

He was showing a general principle. But more recently, Neil Turok has made the point that nowadays it's not a want, it's a *need*. We now have a need for simplicity. Neil Turok has said that measurements made recently suggest the universe is so simple, all of our complicated mathematics can't explain it. He has also said that this means we need new ideas, and that the challenge is to get to the simplicity.

He's worth quoting in detail on this. Neil Turok was director of the Perimeter Institute in Canada for ten years, as he was re-elected for a second five year term. In a way, that's the top job in physics. He won an award for 'leadership in physics' as well, so his opinion counts. In the following quotes he seems quite critical of some present approaches. I was lucky enough to talk to him for two hours at Cambridge University recently, about my interpretation for quantum mechanics among other things, when he was visiting his old place of work, where he held the Chair of Mathematical Physics for a decade. The conversation was filmed for a documentary that was being made.

In one lecture at the Perimeter Institute, called 'The astonishing simplicity of everything', Neil Turok said: *"Simple concepts are concepts which unify. They bring together disparate ideas and disparate knowledge, and make sense of them. [..] Simple concepts are ones which allow us to explain the most we possibly can, from the least possible number of assumptions"*.

And in a filmed discussion with a panel of physicists, he talked about recent measurements (recent in 2014), and what they imply about the universe:

"What's really dramatic over the last three years are the observations - we've discovered at the Large Hadron Collider that there is a Higgs boson. But all of the predictions of supersymmetry, and supersymmetric particles, which are associated with string theory, all of those have not proven correct, there are no other particles so far. And so all the... you know, the most famous particle physicists [...] who were confidently predicting the Large Hadron Collider would see supersymmetric particles, have so far got egg on their face. Nature has somehow found a simpler way. And it's a way which is so simple that we can't yet make sense of it. Nature has found out a way of just having one Higgs boson, and nothing else along with it.

And then we come to the Planck satellite, and again there was this plethora of theoretical models, which could predict almost any pattern on the sky you could imagine [for the CMB], and the Planck satellite measured the pattern last year, and they find out it's the simplest possible pattern. Just describable in two numbers. And the universe is telling us it is simple, it is astonishingly simple.

Yet here we are with thousands of physicists worrying about a multiverse, of infinitely complex universes, which we can't see, and so on and so forth - I think it's great, because they're all on the wrong lines! ...and all these very

very smart people are distracted with a lot of confusing ideas. Whereas I think that... I personally believe we are on the threshold of something really fundamental, which is that we now have the clues, what we need are the theoretical ideas which will resolve these conundra...."

Those words speak for themselves, and are similar to some conclusions that I came to in the mid 1990s, which set me on my way. They show that the real way forward in physics is not the increasing complexity in the mathematical descriptions, that we have at present. It's like the epicycles in ancient science that kept getting added on to correct the orbits of planets, before people knew that they were elliptical, and centred on the Sun. When it arrived, this new approach provided a *simplifying conceptual framework*, and threw out the complicated fix-ups that preceded it. That's exactly what we need now.

So in the present situation, instead of increasingly complex mathematics, the way forward is a re-examination of the conceptual foundations of what we already have. And we need to search for some simple yet subtle concepts, which as Neil Turok has pointed out, the universe is telling us must be sitting there, somewhere just beneath its crusty exterior.

18. The speed of gravity

As you can tell from what I've been saying, later in the book I hope to reveal general relativity's missing twin. I've been hinting at it throughout. But if you actually find the missing twin, it's hard to convince anyone. Carl Sagan said 'Extraordinary claims require extraordinary evidence', drawing on an earlier version of the saying from Laplace. Well that we have, it's a near-proof, it starts on p 95, Chapter 61 - I just wanted to mention it.

I'll end this section by looking at a question about the speed at which gravity acts on things. Does the effect of gravity take time to reach the Earth from the Sun? The speed of gravity is said to be among the most discussed physics issues on the internet, and has caused some confusion. Setting out a few of the questions gives a good overview of the puzzle, but this chapter doesn't reach conclusions. For now I'll just list a few questions and clues.

Newton preferred the idea that gravity has a finite speed, but only making it act instantly gave the right orbits. So Newton's theory has instant action at a distance, and the orbits it produces are so good we still use them. Einstein's theory does even better with the orbits, but it officially has gravity travelling at c, the speed limit for any influence in relativity. We use both theories all the time, and on the face of it they disagree on something very fundamental, but in practice both behave as if gravity works instantly.

We know that the Earth is pulled towards the Sun's present position, and not its position eight minutes ago, which is the light travel time. During the eight minutes, the Earth's position in relation to the Sun shifts, but gravity seems to know where the Sun is 'now', and ignores the delay.

Light doesn't ignore it. Sunlight and starlight arrive from a different direction from that of the source. The Earth moves through space, and it hits the light at an angle, just as raindrops hit a running horse at an angle. This principle, aberration, seems to rule out anything being *emitted* to cause gravity. On the face of it, any emission at a finite speed such as *c* would show up. That is, if the force is directed along the emission path.

The apparent action at a distance isn't a major problem for general relativity. There's a kind of quirk in the mathematics, relating to relative motion, which means that the difference is cancelled. It puts the direction of gravity almost exactly back where it was, compensating for the delay. So general relativity's speed limit for influences travelling through space isn't exceeded, but gravity still acts as if its effect is instant, as Newton and others found out. This point isn't to be questioned in itself - general relativity has an amazing aptitude for being right about things, and it's almost certainly right on the end result of this. But it could still be a parallel set of concepts to the actual picture, and what's really going on is still an open question.

The curved space picture is clearly right or nearly right: the world behaves in a closely similar way. Can we believe that the apparently instant element just happens to emerge like that, from a mathematical quirk? The answer is that if general relativity is right, then yes, definitely. It might be a coincidence that the mathematical quirk appears to give gravity both an instant aspect and a lightspeed one. If general relativity is right, that's what it would be.

But in allowing these two aspects to co-exist, it does something that could be significant - if general relativity is a mathematical parallel to the true picture. After all, the universe seems to have *both* an instant aspect and a lightspeed aspect in another area - quantum theory. Non-local connections were given strong support in 2015, coming from experimental evidence, in loophole-free tests. The links that were shown to exist are instantaneous, but as we have no explanation, the jury is still out on what's happening. I'm not saying the same mechanism is necessarily at work with gravity, but the jury should still be out on that question as well.

But there's a simple principle that does the same thing. What if the cause of the Sun's gravity is already out near the Earth somehow? Nothing can travel faster than lightspeed (in almost any area of physics), but there's no rule that says something can't *already be* somewhere. If the cause is already in place, no action at a distance is needed.

So perhaps gravity doesn't have to travel from the Sun in the eight minutes, because it covered that distance back when the Sun was forming. The Earth feels it locally because the Sun has somehow affected the space around it in a lasting way. If so, the action at a distance (that worried Newton) may not happen at all. Both space curvature and the gravity mechanism in this book are possible candidates for this.

But there are still questions about how the field updates itself. And here the two aspects of gravity (instantaneous and lightspeed) may attach to different areas. The lightspeed part seems to be about changes to the field's shape, the instantaneous part is about directions in the field. The field updates itself at lightspeed if its shape gets changed. Information of that kind, gravitational waves, take time to drift outwards. Astronomers confirmed that their speed is c, or very close to it, in 2017. But curvature, or whatever else is happening locally, seems to tell objects their immediate position in relation to *directions* in the field.

So there are different possibilities. Because the field is symmetrical in certain ways, even if it's being updated eight minutes behind, it might pull towards the Sun's present position. The Sun also has 'midriff bulge', and is not quite spherical, which removes part of the symmetry. But then the planets are all nearly on the same plane, so that might make no difference.

When we get to the flyby anomaly, I'll mention two separate theories, which both do the same thing. Both manage to remove the anomaly, and explain some of the mystery, by putting in what's called 'retardation'. That means a short range *time delay* to the effect of gravity, exactly as if very near to the mass, gravity travels at lightspeed. But if we try putting in a long range delay, the solar system would collapse, so we don't.

Anyway, there you have some of the clues and possibilities surrounding what Newton found - that orbits only work if there's no time lag. The two gravity theories we use, Newton's and general relativity, deal with this in entirely different ways. We have something of an explanation, but questions remain, and our understanding in that area is almost certainly incomplete.

Part 5. The beginnings of a picture

19. A new set of concepts

From here on I'll set out the view of gravity this book's about. The first book, The Unsolved Puzzle, is about quantum mechanics. All three books are about new solutions for unsolved puzzles in physics. They all come out of the same background theory, which is a way to describe space, light and matter at a very small scale. It's a simple enough picture, but it's a lateral, unexpected one. The background theory (Book III) led to two spinoff theories, which are called DQM (quantum mechanics, Book I), and PSG (gravity, Book II).

Book I was on DQM, dimensional quantum mechanics. A documentary was made about it, *The Interactions Avenue,* with two well known physicists and myself discussing DQM and the surrounding area - it can be watched online.

Book II, this book, is about the gravity part of the theory, PSG, or Planck scale gravity. Book I had no mathematics, while this book has. The gravity theory also has a visually interesting set of concepts, it's supported by some simple mathematics, for those who want to look at the equations. But the book will hopefully be of interest whether one happens to love or hate mathematics, as people often do.

The journal paper has most of the mathematics, it's called *Testing a Planck scale mechanism by applying to matter a law for light: a new gravity theory that closely mimics standard theory.* So you might ask why some of it is also here. It's because unlike most theories, this theory has a visual picture that's extremely important to it, and the paper only has it in outline. That's a very bare bones version, for a theory with such emphasis on the conceptual side. Some of the ideas and mathematics from PSG are explained differently in the book, and in a more accessible and complete way. I felt the full concepts and the mathematics should be in the same place. So for some, the whole thing will then come together. Others can skim and skip through the mathematics section.

The end result is that all three books are for people at a wide range of levels of understanding, and I think it works well in that way, for this particular set of ideas. It's a surprisingly simple solution, and hopefully both physicists and a wide range of other readers will find here what I said - three pages back - is needed in physics now: a simplifying conceptual framework.

20. The backdrop

There are many hints that the Planck scale is where all the different kinds of matter comes together. That's where wood and metal are the same, so they fall in the same way. They also respond to other forces in the same way, and there are clues about energy that unify matter - something very *standardised* seems to be going on, in which all matter is the same. So we need to search for some unexpected setup which can do that.

Some will remember (perhaps from Book I or the film) the background arena in which this solution is set. The backdrop looks like a well known picture of the Planck scale from string theory, but it's different in some ways. You have three 'flat' dimensions, making straight lines at right angles - the three basic dimensions. And it's also thought that some small circular dimensions exist, which have curled up all the way to near the Planck scale, where they settle into circular and cylindrical structures. Apart from that, they're just like the other dimensions. If we could see down to that scale, and if dimensions were visible we'd find space has a regular grain to it, with row upon row of parallel cylinders, all pointing in the same direction.

Since the 1980s, when string theory appeared, many have thought space is something like that, and the dimensions are often taken literally. Before that they were seen more as a man-made device to help with our mathematics, like a co-ordinate system.

But now we think they're real. The dimensions are not at all like the idea of the physical world we're used to, but we now suddenly think they exist. That creates a new opportunity to use them in an explanation, and there's good reason to look for one. Their nature may explain the weirdness we often find in physics. We mustn't fall into the trap of assuming that this 'weirdness' *necessarily* needs an explanation, but it still might have one.

In this theory the dimensions make the universe what it is, by vibrating. Light and matter are waves in the cylinders' fabric, or ripples in their surface. The fabric of space can shift its surface a bit, and transmit waves. Matter is waves that travel around the cylinders, light is waves that travel in a direction along their length. A light wave ripples through many of them.

But the cylinders don't have fixed positions. They point in the same direction, but it could be any direction. To picture this, if some people move a chest of drawers from one room to another, they might put it down to rest on the way, at any angle. Its length, breadth and height can point in any direction, but they stay at right angles to each other. The structure of the dimensions is a bit like that, you can put it anywhere.

The mystery of quantum mechanics has a solution in Book I. Some might find a quick summary helpful, to give the picture so far. So the orientation of the cylinders is undecided - there's a set of possibilities. Matter lives on them, as disturbances in their fabric. So this undecidedness affects matter.

We know from quantum mechanics that a particle can be in many places at once. That's unexplained. Matter seems to have a set of possibilities about it, until one gets picked out. In this theory, before a position for the dimensions has been selected, the axes are at many angles at once.

So you get wavelike behaviour: a particle moves along many paths, all set at different angles. That makes a wave. Each path is a possible position for the axis the particle travels along. We know an emitted photon spreads out into a wave. In this picture, that's because it travels all these paths at once.

We know a quantum wave can only be broken up into fixed and equal units of energy. Quantisation is unexplained, but in this theory, the wave consists of many different versions of the same particle. If that's what the wave really is, the different elements in it would naturally all have fixed, equal amounts of energy. But we also know this fixed amount of energy varies from wave to wave. In DQM that's because a different wave will be based on a different particle, seen many times, so the unit of energy can be different.

That's a potted version of the backdrop. Hopefully it gives an overview of the ideas so far (it's set out in more detail on p180). It uses this superposition of the orientation of the dimensional axes - which is standard, basic physics - to explain the only other superposition we know of in this world: the totally unexplained one we find in quantum mechanics.

From here on I'll add to that picture, in a different and simpler area of it, and show gravity at work. Then I'll give a kind of proof, in a straightforward way, that the world at that scale really does look like that.

21. Zooming in on a mass

In the picture so far, there's one thing that might look like a coincidence, but it probably isn't. In the '80s, people got to the idea that the extra dimensions might have curled up rapidly, like rolls of wallpaper, a fraction of a second after the big bang. In the theory here, the scale at which the dimensions start to become 'wobbly', so their skin can easily transmit waves, is the same scale as the one to which they would have curled up.

Surely that's not a coincidence? In a way it doesn't matter: a concept unique to the theory is that the dimensions can transmit waves at the Planck scale, for whatever reason. But it may well not be a coincidence. Perhaps the extra dimensions ended up at the Planck scale in the first place for the same sort of reason, because that's the scale at which they become elastic enough to alter their shape significantly, curling all the way around on themselves and forming into new structures. And the same kind of properties are needed for transmitting waves - enough to create matter.

Either way, when one of the dimensional tubes transmits a wave, a *direction in space itself* will shift its angle slightly as the wave goes by, and then quickly return to its original position.

So a mass like a planet, an object made of matter, is a place where there are many small circles of waves travelling around the tubes. Each looks like the closed string in string theory, which is a small circular loop of matter, usually thought to be made of 'stuff' of some kind. But in Planck scale gravity (PSG), the basic unit of matter is not 'made of' anything - it exists as a disturbance in the fabric of the dimensions, and nothing else.

So a mass is a place where there's a lot of disturbance in the dimensions, and many rotating waves. If you could zoom in on a mass like a planet, at a very small scale space itself would be juddering and vibrating, with the cylinders wobbling here and there, as the waves travel around them.

22. Where sparks fly

This happens across the whole mass, and it led to an explanation for gravity. The idea was that there's a side effect. All that vibration, coming from these rotating patterns within the mass, sends out smaller waves. They're weaker than the circling waves we call matter, and they travel as light does.

These waves are caused by other waves, so they're secondary waves. Where you have a clump of matter in one place, such as a planet, the secondaries start inside the mass and are radiated out from it. They exist at an extremely small scale, so they can't be detected directly. But they come in such huge numbers that together they have a significant effect.

So the planet looks like a sparkler, the hand held firework. There's emission flying out from it in all directions, which dissipates, and steadily weakens as it gets further away, drifting off into space. The further from the mass you go, the less space is vibrating. But near the mass the vibration is strong, and it slows *everything* down. Nothing escapes the secondary vibrations, because they affect the basic velocity at which space can transmit waves.

At a given distance from a mass the secondaries slow everything by the same factor - light trying to get through, matter trying to get through, even other secondary waves trying to get through (which is important later, as it affects black holes). The nearer the mass you go, the stronger those little waves are, slowing the rate at which anything can get through space.

23. A moving refractive medium

Standing back from the picture, there's this emission coming out of the mass in all directions, and spreading outwards. The secondaries travel at the local speed of light, which is a number slightly below c in ordinary weak gravity. At any given point in the field their strength remains constant.

So although there's outward motion happening, the field still keeps the same form and structure. As new vibrations keep arriving at any point in space, the 'density' of vibrations there stays the same. So the effect of the medium on anything travelling through it remains the same at each point, and the field maintains its state, behaving almost like a motionless solid object.

Back with the sparkler, it's much the same. There's motion if you see it from close range, but seen from further away it's like a luminous blob that fades outwards, keeping its overall shape.

The emission dissipates, just as waves on Earth dissipate. It weakens steadily, thinning out slowly as it moves away. This provides a natural explanation for the well-known graded nature of a gravity field, with a long list of effects and aspects of the field reducing in strength in the radial direction. Space itself is what's vibrating - that's also explained (but this time it's just one part of the theory explaining another). Space vibrates because the mass itself is caused by vibrations in space, in the first place. Some will be transmitted beyond the mass: the original idea for the secondaries came from saying one day 'well maybe some of that vibration leaks out beyond the edge of the mass'.

And on Earth waves can slow other waves down, so the secondaries can slow light and other waves, and that part of the mechanism is also given support by what we know. The end result is an interlocking set of explanations, which have supporting analogies from elsewhere.

24. How flat space mimics curved space

It turned out that the secondaries could do several things that make the field work, and do what it does. A medium that reduces in strength as you move away from the mass behaves like the curved 'crater' in space in relativity, the sides of which get steeper nearer the mass. In that picture space is curved at a large scale, but in the dissipating emission picture, space is only curved at a very small scale, to make the circular dimensions.

And you get the same effects. In both, the field deflects passing light, pulling its path around the mass, then sending it zooming away at a different angle. And in both pictures the field effectively slows down a light beam, delaying it, and temporarily changing its wavelength.

Over the 20th century, many good physicists have thought a mass might have a graded refractive medium around it. The approach always looked good, but it had problems. Altering some background concepts might fix them. It leads to what behaves like curvature: for a very loose analogy, space has 'shading' drawn on it, which can give the impression that it's curved.

One can get to a mathematical description of what happens in the field. How light moves, how matter moves, how the secondary waves thin out. It gives the same results to eight decimal places, but the equations are different, and there are measurable differences as well. The refractive medium idea has been incomplete in the past, and short on explanation. It's often about trying to get the mathematics to fit, without much of a picture.

They also didn't get the actual mechanism for gravity in this theory, which is the main part of it, and comes later. But refractive medium gravity has a long and interesting history, which I didn't know about early on. There's more or less nothing to read about it online, which is surprising. There are papers on individual theories, but nothing on the general topic. But there certainly is a general topic, as you'll see in the next chapter. It has been somewhat edited out of human culture (not a conspiracy of course, just that the emphasis on sources like Wikipedia is set by people who have invested a decade or two of study into general relativity, and so don't feel particularly like mentioning its promising rival). They do talk about VSL theories, an overlapping topic, but the heading doesn't mention the conceptual alternative to curvature, which is important. So it's left to me to tell the story.

Some searches in 2004 led to finding out I wasn't alone in thinking some kind of refractive medium makes a gravity field. I was excited and glad to find this out in some ways. It took away some of the loneliness of the struggle, and confirmed some work done up to there. For what it's worth, I reached that idea independently, as many others must have done. Anyway, the idea has a long and respectable history, with some unexpectedly well known characters in it, so it makes a worthwhile story.

Part 6. Refractive medium gravity

25. A suspiciously close analogy

It was Eddington who first pointed out that some effects in general relativity can be interpreted as the results of a refractive medium, as an alternative to curved space. Sir Arthur Eddington was one of the most influential physicists in Britain, and the world, when general relativity arrived. He taught himself the mathematics before just about everyone else, and was very important in waking people up to it.

He was also a celebrated astronomer, and was given recognition by the Royal Astronomical Society, and the Royal Society. It was Eddington who led one of the expeditions to photograph a solar eclipse, which proved Einstein right about the bending of starlight around the Sun, and helped to get his general theory accepted worldwide.

Eddington wrote a book about general relativity which came out in 1920, five years after Einstein's paper. In it he pointed out that curved space isn't the only way to explain a gravity field, and that the effects in general relativity could also be explained by a graded 'refracting medium' that slows light near a mass: '*We can thus imitate the gravitational effect on light precisely, if we imagine the space round the sun filled with a refracting medium which gives the appropriate velocity of light.*'

Two effects for which Eddington pointed this out were the deflection of light around the Sun, and the gravitational redshift. The deflection of light near a mass is identical to the deflection caused by the refraction on Earth, if the medium is graded, getting thinner in one direction.

Eddington lived until 1944, but in the 1960s a third effect was added to the list - the time delay of light. Neither he nor Einstein had thought of it, but by the '60s the technology to measure it existed, and a prediction was made from relativity for the delay on a radar signal bounced off Mars, passing near the Sun on the way out and back. Radar ranging later confirmed it (this is in a later chapter). The measured delay fitted general relativity, but it also fitted with refractive medium theories, in which the medium slows the light near the Sun. The alternative interpretation, though it had problems, had an odd habit of fitting with things.

The similarity to the behaviour of a refractive medium has been mentioned many times since Eddington pointed it out. Not many take it literally, but the mathematical similarities are so striking that it's often touched on, including by the top NASA gravity team headed by John Anderson, in a 2004 paper about the Cassini space probe. By then there were three effects rather than two, and Anderson mentions them all:

'According to the theory of General Relativity (GRT), a light ray propagating in the Sun's gravitational field is effectively refracted, with the vacuum index of refraction n augmented by a refractivity inversely proportional to the distance r from the Sun's center of mass. The gravitational field causes ray bending towards the Sun's centre, one of the classical GRT tests, and in addition, electromagnetic waves are time delayed and frequency shifted.'

We know the mathematics of refraction well: it often fits. It's usually taken as a close analogy, as Anderson does, because the literal idea has problems. Unless they can be removed, it may remain nothing more than an interesting parallel. Roger Penrose also mentions it:

'...one might think of the gravitational field as providing an all-pervading refracting medium, which affects the behaviour not only of actual light, but also of all material particles and signals. Indeed, this sort of description of the effects of gravity has often been attempted, and it works to some degree. However, it is not an altogether satisfactory description and, in certain important respects, gives a seriously misleading picture of general relativity.'

But although for now the approach is flawed, some are aware that changes to the background concepts might lead to a new, viable gravity theory, which could replace - not reinterpret - general relativity.

In 2008 a fourth effect was added to the list. A paper of mine, from PSG, was published on the geodetic effect, which is a very slight tilting of gyroscopes in orbit. This had been predicted from general relativity, but it turned out to have an alternative, parallel explanation.

Gyroscopes keep their orientation angle, unless something very fundamental is shifting them. Measurements from a space probe orbiting the Earth, with four gyroscopes onboard, showed that they slowly tilt through a particular angle while orbiting, just as had been predicted. It was thought curved space is the only explanation for this steady change of angle, so the measurement seemed to confirm general relativity and curved space.

But a few months after that, the paper showed a clear proof that a refractive medium could be shifting the gyroscopes instead. The PSG equation gives the same angle per orbit to 8 decimal places, but is mathematically different. There's a chapter about that later, but it's worth saying here that it added a fourth effect to the list of things the refractive medium picture could explain, and extended it to cover the geodetic effect. That means it extended it to do matter as well as light, which was a step beyond where refractive medium theories were thought to be able to go easily.

Back in the early 20th century the refractive medium interpretation, although it had been pointed out by Eddington, was not taken up by the mainstream physics world. General relativity worked better as it was.

But it was Newton who first came up with the idea, with some work that was *way* ahead of its time, a few centuries earlier. He thought the aether was this medium, but the laws of refraction hadn't been worked through fully at the time, and he had things the wrong way round, so he thought something was displacing the aether near the mass. It was simply too early to get there, but that didn't stop Newton from getting very near to it anyway.

And even Einstein had gone there. In 1911, during a transitional phase while he was trying to splice special relativity and gravity, he published a paper with an equation for a varying speed of light in a gravitational field, with light getting slowed via an effect that increased as one approaches the mass. The slowing he had was *far* more than the tiny one caused by the deflection of light. He didn't necessarily take it in that way, but it implied something that behaved exactly like a graded refractive medium.

So as you can see, the idea has never been too far away, and some very good physicists have tried it out and tinkered with it. No-one is denying that the field behaves like a graded refractive medium. It's just that so far it has been very hard to work with the idea that it actually is one.

26. Problems with refractive medium theories

There have been quite a few RM theories (as I'll call them now), mostly from the 20th century. What I'll say here doesn't necessarily apply to all of them, but the idea has had problems, and some of those I can remove.

The only piece I've found about RM gravity generally, was a paper showing just RM theories of one kind. It's on attempts to interpret general relativity, rather than to replace it. Ever since Eddington in 1920, people have tried to explain Einstein's theory via refraction instead of curvature. Some try to find the refractive index for bits of mathematics such as the Schwartzchild metric, to see what it would be if refraction is the cause.

But that's trying to insert RM gravity later on, when the mathematics had already got complicated. Sometimes the refractive index they derived would be many numbers added together (a convergent series). But if one wanted to *replace* general relativity, instead of interpreting it, the need is to insert the RM earlier in the sequence, to get a simplifying conceptual basis underneath the picture. If it can apply to both light and matter, the mathematics might reach the simplicity that many think we now need to find. And as we'll see in the geodetic effect, the simple little refractive index from the mathematics of PSG may already be doing better than these complicated approaches: it was upheld in 2008, as an alternative to general relativity, by experiment. So it looks like that might be the refractive index, if there is one.

Now we go back to the story. In the early 20th century, RM theories were at a disadvantage. They seemed to hit exactly the same problems that the aether had run into, which included tough issues like the medium not existing. We were far from keen on the aether by then.

Before that time people had invented other substances to perform particular functions: two of them, phlogiston and caloric, had been created to explain combustion and heat transfer respectively. Both had turned out not to exist, and now we had been forced to shake off the aether as well. So people were getting fed up with inventing imaginary substances, making up stupid names for them, and then finding out that they didn't exist.

In fact, people threw out more than the bathwater when they threw out the aether. All that had been disproved was a transmitting medium that behaved like matter. But perhaps somewhat 'on the rebound', physicists threw out all possible transmitting mediums. Later, in the 1980s, things shifted because of string theory, and the dimensions started being taken literally. They also now seemed to have structure at a small scale. But by then we had a history of no transmitting mediums (not material ones - only what we call fields, which are largely unexplained). So in the late 20th century, comparatively few saw the new possibility that had quietly landed in the arena.

But back in the early 20th century, for RM theories this doubt about mediums was part of a wider problem. It was about what the medium actually was. It had to be more dense near the mass, and less dense further away: for some reason it crowded around the mass. But one couldn't say it did this because of gravity - the medium's distribution should *cause* gravity in the first place. So some other reason for its gradedness was needed, and that was often missing. And in general RM theories have had difficulty defining the medium. That is, 20th century ones. It was easier for Newton, as in his time the aether seemed to provide it.

In PSG, most of these problems go away. The graded nature of the medium is explained as the result of the dissipation of emission. The existence of this emission is explained from within the theory, because it takes the form of vibration, and the mass itself is caused by other vibration. And the medium's undetectability is simply due to the very small scale.

But another question is what fabric, or underlying medium, vibrates to cause these waves. The waves themselves create a graded medium, at the next level up, but being waves, they have to be 'in' something. I've mentioned the advantages of the curled-up dimensions for that, a good one being that it's not an invented medium. The fabric of the small-scale dimensions is widely thought to exist anyway nowadays, for different reasons.

And it's thought that space can vibrate. Space seems to transmit light: waves of all sizes travel through it at the same velocity, just as waves of all sizes on Earth travel though a given medium at a fixed, characteristic velocity. When applied to space, this has been an incomplete idea. But many believe it, and lists of reasons in Books I and III support it very strongly (including reasons from relativity, such as the fact that space transmits gravitational waves at lightspeed). Because space transmits waves, it's reasonable to build a theory in which the dimensions of space can as well.

The dimensions being taken literally greatly helps the PSG picture. It removes what would otherwise be a massive assumption. It uses, and depends on, a medium people think exists anyway. So the whole set of ideas is economical in assumptions: it assumes less and explains more. It means that Eddington's 'refracting medium' can shed some cumbersome assumptions, in the light of more recent developments.

27. A rare medium

To make these advantages possible, the medium in PSG isn't a standard one. It's not 'stuff' that fills the space around a mass, it's vibrations in the fabric of the space there. This makes PSG different from many theories. One question is whether vibrations could cause refraction at all.

It has been known for centuries that light travels slower in glass, water or air than in a vacuum. Any material transparent enough to let light through is still more opaque than space, and light will slow down in it. In glass, light travels at 65% of its normal speed, and in water, 75%. This helps RM theories, as it's known to happen beyond any doubt.

Some 20th century RM theories implied a conventional medium of that kind, behaving as materials on Earth do. But vibrations can also slow waves down, also on Earth. If a bit of metal is vibrating, sound will travel slower through it than the basic speed for sound through that metal. Waves in a material can reduce its transmission speed. The same happens with water waves - to put it bluntly, waves sometimes slow other waves down.

And if you go to a small enough scale, that principle might be the underlying cause of *all* refraction. According to the background theory, the reason light travels slower in glass than in space is same reason that sound travels slower through vibrating metal than through still metal. Matter is really nothing but a lot of regular vibration, so a sheet of glass is just a space that's vibrating in a particular way at the Planck scale. This slows other vibrations trying to get through that space, such as light. So light's speed goes down by a different factor for each material, as we know it does, because each has a different pattern of vibration going on inside it.

All you ever have is space - sometimes vibrating space. But matter, such as a sheet of glass, is an ordered version of that, because it has circular vibrations in it, and they're very even. But around a mass you have a less regular kind of vibrating space, the secondary vibrations pouring outwards. But the principle is exactly the same, and the secondaries also cause refraction.

28. What happens with matter?

But the main single problem with the whole RM approach is about explaining the behaviour of matter. In a gravity field, for instance in orbit, objects made of matter do things they never seem to do on Earth.

By the mid 20th century, alongside a new picture of gravity, we were tracking objects in the solar system very accurately. By the '60s, radar signals could be bounced off planets, returning to Earth to give the planet's exact distance at a given point on its orbit. So the whole orbit could be traced and mapped, and compared with predictions from theory.

People found that small corrections to Newton's orbits came out of Einstein's theory. According to general relativity, the relativistic adjustments are mostly the result of space being curved. A mathematical representation of gravity can be arrived at by taking Newton's equations, and then adding in the effect of space curvature. This gave a better approximation.

The resulting differences are small, because Newton got it very nearly right. According to general relativity, he did so because space is very nearly flat and only slightly curved. But the adjustments, when added in, still improved the orbits, and made them fit measurements better. So space curvature looked unavoidable. In fact, adding a RM instead of curvature does the same thing mathematically. Both must be added in after Newtonian gravity, separately: in PSG this is for specific reasons, as I'll explain.

But there was a problem with matter. In Newton's gravity theory, objects on elliptical orbits accelerate as they approach the mass, then flip by it quickly, slowing as they move away. But on top of this pattern of speed changes from Newton, in the 20th century a *far* smaller reverse pattern was predicted, then found. Curvture adjustments meant orbiting objects slow down slightly (from their expected speed) as they approach the mass and enter the strong part of the field, then speed up slightly as they move away. With matter, this had to be disentangled from the Newton orbit, but light does it in a more obvious way, slowing down, then speeding back up again.

The easiest way to imagine this, with curved space, is to say speeds seem to change because we only see what happens in three dimensions. In four, light and matter keep their real speeds, but have further to travel, because of the extra distance along the crater floor. So if either light or matter approaches a mass, it starts to enter something like an invisible crater, slowing down from its expected speed because it has further to travel than we see, around the curve. Then, moving out of the strong part of the field, it speeds back up again, because we see more of its motion.

This picture of general relativity is one of several equivalent pictures. It's not always relevant, although it fits the mathematics. The fourth axis is thought to be time, and both space and time are curved, which complicates things. But it's a good picture, for those who look at the conceptual side: it shows a way for light to keep a constant speed, and yet get delayed, and we know it does get delayed. And though to some the mathematics is all that's needed, many think the true picture is something like that.

Whatever the picture may be, these small differences have been measured, and the radar ranging data works better with the calculated orbits when you put in curvature - or whatever else behaves like it.

Now when RM theories step in and try to deal with these adjustments, they have no problem with light. RM theories go by what we observe directly, and depend less on unobservables - ideas. To a believer in general relativity light approaching a mass "slows down", but to a believer in RM gravity it simply slows down - without the inverted commas - as it does in water or glass.

Both theories are happy with light doing that, and they're also fine with the fact that light speeds up on its way out of the strong part of the field. In RM theories the behaviour of light is well backed up by what happens on Earth, where light is literally slowed down.

It also speeds back up again. Shine a torch through air, into a tank of water, then out the other side. It slows down in the water, due to a higher density, then regains its earlier speed out in the air. This is well known - light likes to travel at its normal speed, or as near to it as possible. If anything impedes it, it slows down, but only for as long as it has to. So light passing close to the Sun, where we know it gets delayed, could be passing through an area with a higher density RM, then speeding back up as it moves away.

But what *is* a problem for RM theories, and it's a major one, is that they can do light, but they can't do matter. Apart from in quantum effects, matter is often not seen as being affected by refraction at all.

But because near a mass matter slows down just as light does, it's interesting to ask whether a refractive medium could be slowing both. As I said, both are slowed by the same factor at a given point in space (from the speed at which they would otherwise travel), although general relativity usually expresses it differently. Many physicists are aware of the similarity, and to many it looks like curved space is affecting both in the same way.

But if you say instead that some kind of refractive medium is affecting both light and matter - and people who believe in RM theories for light can't really ignore the similarity in how they both behave - you then have to explain why matter speeds back up again, whenever it moves out of the stronger gravity region, near the mass. And that's the problem.

Matter here on Earth, you see, doesn't speed back up at all. Out in the field, I've seen this principle at work. Kick a football through some long grass, and when it comes out on the other side, it will be moving slower. It will have lost some momentum, and emerging where there's short grass certainly won't make it speed back up again.

But in space, planets on elliptical orbits go by these small adjustments which, if disentangled from Newton's orbits, involve speeding back up after closest approach. So perhaps general relativity is right, and curved space affects light and matter in the same way. Or alternatively - and this produces exactly the right mathematics - if a refractive medium is at work, then at the scale of this medium, matter would have to be more like light than we think.

Part 7. A simple mechanism

29. A law of nature lost and found

I haven't told you the main mechanism of the gravity theory, which to me is the most interesting part. So let's return to the apple in the tree, just before it falls, and ask what happens at a small scale. At present there's no agreed explanation for why an object released in the field starts to fall. So when the apple breaks from the stem, we're not sure what's happening.

The refraction of light is well understood. Light moving from one medium to another changes its speed and angle. The change of angle is given by Snell's law, which was found in the 17th century by Snellius in Holland, but the law had been set out centuries earlier in different forms, or ideas approaching it. Ptolomy, in the 2nd century, got to at an incorrect version of the law, and he was then either mistaken or naughty, because he presented his numbers as experimental results, rather than calculations. Knowing the correct law now, we know they can't have been from accurate measurements.

Ptolemy's false version drifted around for eight hundred years, and was then perpetuated in a 10th century translation of his work by the Basra scholar and major physicist Alhazan, which kept it in circulation for another six hundred years. But ironically Alhazan also translated another book, from nearby him in time and location, with the correct law. The author of that book, written in 984, and the man nowadays seen as the discoverer, was Ibn Sahl, a physicist and lens maker from Baghdad. In it he said Ptolemy's version was wrong, and showed how the real law works, with a geometrical drawing.

Six hundred years later, when exciting new scientific progress was sweeping Europe, the law was rediscovered several times (whether by examining light or each other's work), including by Snell, Descartes and Harriot, in Holland, France and England. Snell was thought to be the originator for centuries, and the law is still named after him. But since the 1990s, when historical research uncovered a much earlier version, the credit has been back with Sahl. Things finally went his way almost exactly 1000 years after he found the law, so the lesson is about patience and perseverance.

It's a simple law that relates speeds and angles. Say a light beam moves from air into water. Hitting the surface at an angle, it then dives into the water at a steeper angle. This might be unexpected - moving from a thinner medium to a denser one, it's harder for the light to get through. One might expect it to take a shallower angle, almost as if trying to avoid the water.

But it does the opposite, and the reason is well known. The light beam has width at some relevant scale, so when it arrives diagonally at the horizontal boundary of the water, hitting it at an angle, one edge gets there first. That side of the light beam shifts to the slower speed before the other side. So for a short time the upper edge is still travelling through air, and moving faster. This brief mismatch of speeds swings the light around like a lighthouse beam, and it dives down into the water.

The same effect can be illustrated using, of all things, a tractor moving from gravel onto grass. The tractor goes more slowly on the grass, and when it hits the border at an angle, one of the drive wheels gets there before the other. So for a while the tractor is turning through an angle, with one wheel moving faster than the other. This process swings it around, and it heads off more steeply across the slower medium, just as the light beam does.

When Eddington described refraction, he pointed out that a wave reaching the shore, if one side reaches shallow water before the other, can be turned through an angle. All of these examples involve a brief difference in speed across a moving object, creating a change of angle.

The mathematical description is very general. Snell's law doesn't need details like the width of the tractor, the wave, or the light beam - the details happen to make no difference. It just relates speeds and angles, before the turn and after it, in a very general way. So the law of refraction can apply to a wide range of situations, at a wide range of scales. It can happen at any scale at all, and I'm sure you know by now that I think it happens at the Planck scale. Snell's law in its simplest form is

$$(sin\,\theta_1) / (sin\,\theta_2) = S_1 / S_2 \tag{1}$$

θ_1 and θ_2 are angles from the normal ('the normal' is a line at right angles to the boundary between the materials). S_1 and S_2 are the speeds for light in the two materials - this is even simpler than using the refractive index, but it's much the same. The law is that the ratio between the sines of the angles is also the ratio between the speeds.

30. The bouncing laser

There's an interesting refraction effect on Earth, which is well known. I think it's the best clue on Earth about how gravity works. It's about a laser beam sent through a sugar solution. The experiment is done a lot, and is described on educational websites. If some sugar is thrown into a tank of water and left to dissolve and settle overnight, it distributes itself with a density gradient. In the morning you find something like a tequila sunrise, but instead of upward layers with density jumps - grenadine, orange juice, tequila - there's a steady gradient, which also gets thinner going upwards.

It's a steep one: at the bottom of the tank the mixture is around 84% sugar, with a refractive index of 1.5. But a few centimeters up you have pure water, refractive index 1.33. So just a little way up, the speed at which light travels shifts from 2/3 of its normal speed to 3/4 of it. That's a speed difference of 25,000 kilometers per second, across a few centimeters of height.

Then a laser beam is sent through the tank horizontally, and what happens is surprising. Below the beam there's a slower refractive medium, but above it there's a faster one. The beam moves horizontally at first, but its upper edge is travelling faster than its lower edge, which swings it downwards. The beam angles itself downwards for the same reason the tractor turns. But unlike the tractor, which hits an abrupt change between two materials, the laser finds the liquid increases in density steadily. Being in a graded medium, instead of taking a straight path it follows a curve, swinging further and further around towards the vertical.

And rather amazingly, if at the bottom of the tank there's a mirror, the laser will be reflected and bounce off it, curving upwards again. It will then curve around and back on itself, and the beam can be adjusted so it reflects several times off the mirror as it travels through the tank, bouncing away like a stone skimmed across a lake. This behaviour of light is well understood. It's simply due to refraction, and what happens can be predicted using very little more than what we still call Snell's law.

31. Why things fall downwards

So here's the gravity mechanism from PSG. Describing it with an apple might seem a cliché, but whether or not this is the right answer, it might as well be applied to the original question.

So imagine the apple in the tree, seconds before it falls. When you zoom in, the matter in the apple is a large number of circling waves, and in relation to the gravity field, they're making horizontal circles. They travel at the same speed as light waves, and through the same medium, which is space. Near a mass, that medium has a graded refractive index.

Below the apple, many small waves are flying up from the ground, thinning out as they go, all the way up to the branch. The secondary vibrations come out of the Earth and fly upwards through everything. So the apple finds itself in a graded medium. Now let's say the stem holding it to the tree breaks.

General relativity has questions on how the motion gets started, but in PSG the apple is made of spinning waves. Motion is already happening, so it's just a question of what angle the waves will rotate at. Below them, towards the mass, they can only travel more slowly. Above them, away from the mass, they can travel faster.

So being like light, which always angles itself into the slower medium, their paths turn downwards. They get angled slightly into the denser medium, just as in basic refraction, and they start to 'spiral'. The circles become helixes, so the apple starts to fall. The medium's density increases, the waves dive into it at an increasingly steep angle - that is, the apple accelerates.

So zooming in if we could, as the apple falls, its matter looks like a lot of little springs. Trillions of descending 'spirals' are all moving together, in an apple-shaped cloud. As its speed increases, tiny processions of waves, all travelling in parallel, get refracted around tiny tubes made of space. The angle on each helical path gets steeper, each helix gets more stretched out, and the apple falls faster and faster. Then it hits the ground, and dust flies up.

32. A sensitive system

This visual description might or might not be interesting, but the question is whether it gives an account of what actually happens. It's comparatively easy to test it and find out, because the mathematics of refraction and gravity are both well known. The surface on which the refraction happens is 'rolled up', which transforms the shape of the path, and a familiar process is then hard to recognise. But like any lateral jump that's obvious once you know it, the transformation from one to the other is simple, and in the visual picture the unrolled state translates to the rolled up state quite easily.

The same happens with the mathematics. There are many places where the mathematics of refraction 'rolls up' into the mathematics of gravity. Both are well known, but the indirect and almost cypher-like link between them is far from obvious, and remains hidden unless one is looking for it, as I was when I found it. Before testing it, I'll finish describing the picture.

The apple aims towards the centre of the Earth, as that's the direction where the medium increases in density fastest. The laser beam in the sugar solution tries to do the same, and this is well understood. A graded medium will steer light towards a vertical path, where the gradient is steepest.

But a small-scale helical path seeks out the vertical far more efficiently. If it aimed a little to the left or right, the density increases less steeply. The apple is enormously sensitive to these differences - it's responding to them at the smallest scale we know, and the slightest change to the transmission speed affects the array of little 'rotors' inside it.

The refractive medium surrounding a mass is very finely graded, all the way up through the field. It arises from huge numbers of incredibly small waves, and they get averaged over. So you can let go of an object at any height, and it will always find itself on a density slope. And it will notice the slope across very small distances - even the size of an atomic nucleus.

What makes this possible is the scale. The nucleus of an atom (or rather, the space it occupies) is vast compared to the Planck scale. It's larger by a factor of 10^{20}, the factor by which the visible universe is larger than the Earth. So because matter ultimately arises at the Planck scale - and as in string theory, vibrates and zooms around to fill a space like the nucleus of an atom - it can register a slope arising at the Planck scale, across what to us are very small distances.

Earlier versions of RM gravity didn't have helical refraction. They got some of the large-scale picture, but none of the small-scale one. Certainly no-one has discovered the mathematics of helical refraction gravity, because it contains a major surprise, and would be well known if anyone had found it. According to PSG, it makes it possible to complete the picture.

Refraction creates a pull. It pulls light, but we still don't think of it as a force. That's because forces tend to pull matter, not light. But if an ordinary object is really made of spinning light, that's different. An object like that would be pulled by refraction. And the force created would be a real, direct force, that uses $F = ma$, as forces do. So this jigsaw piece is beginning to look the same sort of shape as the hole in the jigsaw we were peering into earlier.

33. The Galileo experiment, but with light and matter

From here on, the exciting places are wherever this picture becomes visible, because it connects with experiment. This chapter is about adding light into a well known experiment with matter.

A paper on the history of physics says: *'Human efforts to understand vision and the nature of light remained at the forefront of intellectual endeavours for well over 3000 years.*' There has been a long, slow journey of discovery, in which our understanding of light kept developing and shifting, and it's not over. Book I includes a new explanation for the wave-particle duality, which might be some food for thought (summarised on p180 here). This chapter is about light and gravity.

When Apollo 15 went to the moon, David Scott took a hammer and a falcon's feather, and dropped them together. With less gravity and no air resistance, they fell slowly side by side, and it was easy to see they hit the ground at the same time. The reference section has a video. Galileo's experiment, done in this interesting new way, was even more immediate. But his original version of it also was (whether or not it actually happened). Another variation on it follows, which shows that it's wider than we thought.

The idea is to release light and matter at the same moment in a gravitational field, from the same height. A light beam is sent off horizontally, and a piece of matter is released to fall vertically. Gravity will affect both: we know that a mass can pull light towards it, bending its path.

So a light beam sent horizontally, from close enough to the Earth's surface, would later hit the ground. The beam has to start very close to the Earth, perhaps a nanoscale type distance above a long, flat metal strip. Otherwise it might just fly off into space, at a tangent to the Earth. Air resistance has to be removed, as with the version done on the moon.

But this is more of a 'thought experiment' than a lab experiment, for now at least. It leads to a calculation: one ignores both the curve of the Earth, and the post-Newtonian adjustments (Chapter 44). The result can be calculated via standard theory or PSG, and they give the same answer, which is that the light and matter hit the ground at the same time.

A small piece of matter is released, and falls towards the mass. It accelerates, its speed increasing up from zero very slightly, but keeping to a slow speed over a tiny distance. The light, meanwhile, keeps to an extremely fast speed throughout, covering a large distance. Make the height they're released from a hundredth of a millimeter, or 0.00001 meters. Matter simply falls that very small distance. In the same period of time, the light travels 428 km, slightly above our hypothetical flat metal sheet, then hits it. So why do they hit the ground at the same moment?

The first point to make is simply that although standard theory and PSG may disagree after many decimal places, the results are closely similar. Both show that this phenomenon exists. And it is a phenomenon, because it works from any height chosen at random: not just above Earth, but any spherical mass, as long as both light and matter reach the ground.

But if you look at the question of why it exists, PSG has a better explanation. I've never seen mention of this elsewhere, and have scoured the internet for traces. There's no reason for anyone to think of it, or do a calculation, unless they're looking for it or expecting it, as I was. But there's a 'reason' for it, of sorts, in standard physics. In fact, it boils down to saying that it's the same as some other unexplained physics.

There may be a different way to explain it. It could have been discovered any time since Newton, because the calculation is done via Newton's theory. But a proper explanation is needed, more than simply showing how it comes out of the mathematics.

So here's one. PSG has a simple reason for it: the path taken by the light is the same as the one taken by waves in the piece of matter. One path 'spirals' around a cylinder, while the other travels almost straight, but both have the same cause - which is, mathematically, standard refraction.

The light and matter both move through a medium that increases in density towards the ground, and the two stay at the same height throughout, as the hammer and the feather did on the moon. One path is 'rolled up', but they're identical in every other way. The paths are exactly the same length, and at any given moment they have exactly the same gradient, exactly the same height off the ground, and are being travelled at exactly the same speed. So they reach the ground at the same time.

The path taken by the light beam is just like the path taken by the laser in the tank. The laser beam descends quicker, but only because the sugar solution changes in density faster than the Earth's refractive medium.

When they're released, the light and matter travel away from each other at right angles. The fact that the paths are closely related makes sense in PSG. The link is about how they travel through the structure of space, which has right angles in it. One path is the other 'unrolled', and looking at the small-scale picture, both set off horizontally. But in standard physics, they don't both set off horizontally. In standard physics, it's far from clear what the link between the light and matter could be. But in PSG it pinpoints the similarity between them, and shows it at work.

34. A box of mirrors

In a variation on this idea, it's possible to make the light go around a circuit. The end result is the same, and you find that you've accidentally built a large-scale working model of what gravity does to matter in PSG.

Say a large cylindrical box is made, with a hexagonal cross-section. There are mirrors on the inside, with equal angles between them. So a light beam can bounce around inside it, from mirror to mirror, taking a hexagonal path of its own (offset by half a side of the hexagon). In principle, light could be inserted into the box, and left to go around many times. If the cylinder is vertical, the light descends due to gravity, 'spiralling' down *with the same acceleration with which an object released at that height falls*. The light would then be behaving as matter does in PSG, but at a much larger scale.

The length of the light path is the same as that of the two paths described in the previous chapter. For the calculation, there'd be a need to allow for the imperfect aspects of it. But the size of the cylinder makes no difference, and although the number of collisions varies between cylinders of different sizes, in all of them the light path to the ground is the same length. This might also be done in an even simpler way, with two parallel mirrors, and light bouncing between them. The hexagonal version imitates the small-scale world better, as it approaches having a circular cross-section.

The result would agree with other versions and calculations on it, including ones that come from standard physics, on the time taken for light or matter to reach the ground. Does this box of mirrors show us light or matter? Light, but in a way it shows both. What it really shows is a well hidden, unexpected similarity between them. And even if it's only in a calculation, the small-scale behaviour of matter in PSG, and the point is, *of gravity*, would be shown to work - via a large-scale simulation.

35. How might standard physics explain this?

Returning to the original point about light and matter released together, it was only noticed at all because PSG has a specific explanation. I'd never have known otherwise. I don't know if any others know about it, perhaps a few do - I've never seen any mention of it anywhere. But a look at the mathematics of projectile motion shows that standard physics includes the same thing. It was Galileo who made the discovery relating to it, though he might not have known that a lamp shone from the tower in Pisa, as he released objects from it (according to legend), could have any bearing on the experiment at all.

But it's not a coincidence that it was Galileo who found this, because he was studying the way that matter falls, including projectile motion. That's about the trajectories of objects that are thrown, or fired off in any direction. Being a very good self-taught physicist - having 'dropped out of college' - and full of an inquisitiveness that had not been blunted by the passive relationship that formal academic training can sometimes be, he discovered the downwards acceleration is independent of the sideways motion. That means that objects thrown horizontally from the same height, which is a simple version of this, will all have the same downward acceleration.

Now light doesn't always go by the same rules as matter. Assuming it always does is a trap that physicists too often fall into, in my view. Still, it sometimes does, and in this area it may. After all, light gets deflected by a mass via the same equation as matter on a hyperbolic orbit (as in Appendix B). And in this question, it's worth trying out the rules for matter on light. You treat light as a particle, as Newton did.

If you say light emitted horizontally is a projectile, applying the mathematics of projectile motion to light, you find that it hits the ground when PSG says it does, but via different equations. With horizontal motion, the equations are rather exceptional. The faster a projectile's speed when fired off horizontally, the greater the travel distance before it hits the ground. But the time it takes on its journey is always the same: you can fire off a projectile at 200 mph, or throw a stone at 3 mph, they'll both take the same time to hit the ground, if they start at the same height.

So yes, light hits the ground at the same moment matter does, working it out using some basic gravity equations. But then, so does any other matter you throw horizontally in that general direction, at any speed.

So is that an explanation? Can we simply say - well, everything does that! We can, but there's a need to explain why. And this is the uniform response of matter again. Galileo experimented on both vertical and projectile motion, and discovered they're linked. In both, the downward motion due to gravity is the same for all matter, and separate from all other motion the object may have. When projectiles start on a horizontal path, this becomes particularly noticeable. And it turns out that light does it as well.

Nowadays, a common explanation for this uniform response is that space is curved near the mass. But I've shown that to be no explanation, because the uniform response of matter is everywhere, not just in gravity, and it always goes by the same equation. This means a wider explanation is needed.

It also means a different explanation for this point about light and matter is needed. There's nothing in standard physics, as far as I know, that doesn't fail to explain it, asking many more questions than it answers. As usual, one way to explain things in standard physics is merely showing the way that it appears out of the mathematics. But meanwhile, the explanation from PSG is there in the chapter before last, raising its eyes to the ceiling, and pretending not to hope that anyone might be interested. Because the 'Light and matter at right angles' thought experiment gives good support to PSG, I've described it here. But the equations, and more about it, are in Chapter 70, which is at the end of the 'Mathematics of gravity' section.

36. The 'pause in the air' point

When someone throws an object up into the air, or kicks a ball upwards, it reaches a point where it pauses, then starts to fall back again. It's quite hard to make a theory that shows why gravity does this, but the laser in the tank also has a turnover point on the arc of its path. in PSG the same thing is happening, and 'unrolled refraction' is the same as helical refraction. Later I'll show how PSG leads to an equation for the height of this turnover point, if you know the initial upward speed. The height it gives is different from the number from standard physics - by a few ten millionths of a meter.

This leads to an interpretation for escape velocity, and explains why it exists. It comes out of the mathematics, but for now we're still on the picture, so I'll describe visually what happens at the pause point. But if a football is used I should make something clear - although a football bouncing away across the ground looks like the bouncing laser, they're not very similar.

The football and the laser bounce differently, but there's a connection. The laser shows what's happening with each small part of the football. Each bit of matter is like a rolled up version of the laser, and has refraction angles, and goes by Snell's law. But the angle of the football's large-scale path is not a refraction angle. It comes out of how matter behaves at a large scale, where the rules are different.

So that angle is irrelevant. Let's remove it, or rather, make it ninety degrees, by just booting the ball upwards vertically. Probably best to kick it as hard as possible. It rises through a graded medium, which is thinning steadily.

The harder the upward kick, the steeper the initial angle on the helical paths, up from the horizontal circular path. And the steeper this angle is, the more stretched out the helical paths in the ball are, so the faster it rises. As is set out in the chapter after next, in PSG a velocity can be defined by an angle. The initial upward velocity is v, and the higher its value, the larger the angle caused by the kick. So the longer that angle takes to be refracted around to the horizontal. The angle gets smaller, and so does the speed.

As the football climbs far above the turf, it rises through a steadily thinning medium. In simple, basic refraction, light entering a slower medium changes direction, taking a steeper angle. Conversely, light entering a faster medium does the opposite, and takes a shallower angle. Accordingly, the path angle gets steadily reduced, and the ball slows down. Then it pauses in mid-air. Its matter briefly makes horizontal circles, but the waves find themselves on a density gradient, so they get refracted downwards again.

As I said, the idea that refraction can have a 'turnover radius' is shown by the bouncing laser. It's not that it reflects off the bottom of the tank. It's that it rises, levels off, and then starts to descend again. I don't know if Ibn Sahl or Willebrord Snell knew that refraction can do this in a graded medium, but the law they discovered can. This hopefully makes the idea that gravity uses that same law more convincing, because objects sent upwards can reach a turnover point, and fall back down again.

37. An action replay

When the ball pauses in mid-air, its waves trace out exact circles. But in fact, they're on nearly circular paths all the time. When a force such as a kick to a football is applied, it will only give the object's matter a tiny shift of angle on the rotation paths, about a millionth of a degree. But that's quite a lot. A ten millionth of a degree leads to a speed of about 1 mph. A change of angle like that will be negligible elsewhere, but matter is enormously sensitive to these changes, as it's rotating at such a high speed. So a millionth of a degree will still be enough to send the ball 'spiralling' upwards at 12 mph.

Standing back to look at the whole match, and trying to see it at a small scale at the same time, there's a need to sacrifice a lot of accuracy. To give an idea of it, linking up two scales, the video will have to be in slow motion, through a very wide-angle lens, and altered in other ways.

But after a few adjustments, what you get is a slowly shifting image: a boot moves towards the football, an arm moves through the air, the ball bounces slowly away. And everything is made of clusters of parallel springs, stretching outwards together, flattening back to circles, shifting and wobbling sideways, drifting and getting refracted downwards again, as objects move and bounce around in relation to each other. It's a glimpse of a strange world, and it's odd that although invisible to us, it may have been hidden there all the time, sitting right underneath the familiar world we see.

Part 8. PSG theory

38. A simple mathematical theory

From here on there will occasionally be equations, to add to the description. Until the mathematics section there aren't many - in the next 30 pages there are three. Most of those are in this chapter, and I'd ask readers who are not mathematically inclined just to skim or skip past them. And the mathematics section still has good stories in and after it - two of the best are about black holes and the mass discrepancy.

I've put all non-essential mathematics towards the end of the book (the post-Newtonian adjustments, and predictions for experimental results) along with anything in-depth, in a section for people who know mathematical physics to some extent at least. But here I'll describe the picture from PSG, and later I'll try to show that it's true. A simple near-proof, or what's known as 'smoking gun' evidence, comes out of applying to matter some mathematics that's normally only applied to light.

I've described a picture of the smallest scale, in which light travels along the cylinders, and matter travels around them. With matter, even when to us an object is stationary, its matter is still travelling around a very small cylinder, which is curled off into a fourth direction.

But if the object moves, the waves in its matter travel in a direction along the length of the cylinders. But they still go on travelling around the cylinders as well, so they move along and around them at the same time. That makes a helical path. We only see the slow, forward part of its motion, but each bit of matter within the object is really travelling at lightspeed all the time, looping rapidly around the cylinder as it goes.

The cylinders are aligned with the direction of motion. What I've described so far is what happens in a particular frame. If we decide an object is moving in a specific way, we're choosing a viewpoint, or reference frame.

Here's a picture that's easier to visualise. Say an object isn't moving, then it starts moving. If it could be seen at a very small scale, it starts as a cloud of rotating circles, then it moves away looking like a cloud of little springs. The object's particles are all 'spiralling' along together on parallel cylinders. Each unit of matter moves in the same way, so it's oddly regular and geometrical.

The overall speed of each helical procession of waves is lightspeed, because that's the transmission speed of space, for waves of all sizes, and here 'space' includes the fabric of the cylinder's surface.

'Spiralling' is not the correct word, but it gives the picture. Mathematically, the overall speed c is made up of two speed components at right angles: the visible, ordinary velocity through 3-space, v, which we see and measure; and the rotational speed component, r, which remains hidden, and is usually much faster. They combine in the way that speed components at right angles do, and the third speed (the overall speed) is lightspeed, so

$$c = \sqrt{v^2 + r^2}$$. (2)

And coming out of the background theory, a single *angle* for the helical path is used to define any velocity for matter. If the cylinder could be unrolled and flattened out, the helical path becomes a straight line. That leads to an angle that can describe any regular helical path, either in standard geometry or PSG. It's like an inclination angle, and in PSG it corresponds to a velocity through ordinary three-dimensional space. The angle θ in degrees, is:

$$\theta = \arcsin{(v/c)}$$. (3)

A velocity of zero m/sec (ie. the object isn't moving) means an angle of zero degrees. If the object moves, each circular path becomes a helix, so the angle θ goes up. θ can be worked out from the velocity v. For an everyday speed it's a tiny angle, because the helical paths are very tightly wound, due to the enormous speed around the cylinder. Walking speed, about 4 mph, gives an angle of 0.0000003, less than a millionth of a degree.

That path angle comes in again later. But for now, this simple structure I've described allows a solution for one problem we've been looking at. Matter's true nature is like light at that scale, so when it travels through a small-scale refractive medium, it responds as light does. When an object enters a slower medium - say it travels through a gravity field, where space has a decreasing transmission speed - all three speeds on its helical paths will be slowed down together, and all by the same factor at any given moment. So they keep their simple relationship, and the path angle θ stays the same. It happens because v and c in equation 3 both change by the same factor.

So equation 2 applies at any point in a gravity field (with c as the local speed of light). As the object crosses the field, the way space is vibrating changes steadily, and so do these speeds on its helical paths. But they always do so in unison, keeping their underlying relationship. And the key point is, the two hidden speeds allow matter's visible speed, v, to do what had always seemed out of the question. They allow that speed to respond to a RM via a simple factor difference, just like we never see matter doing on Earth.

39. Making matter like light

This picture removes the 'speeding back up' problem, which is about orbits. But it also removes the other, much larger problem, which has always made RM gravity seem less than credible. In RM gravity, as I said, it was hard to see why matter should be slowed by a straight factor difference.

In the world we know, if you try it, matter's velocity will follow a much more complicated equation. Slow matter down, and you tend to get a drag force: in many situations it will steadily decelerate, in a cumulative way. But if it's to respond to a RM as light does, it's just slowed by a fixed factor - no drag force, no cumulative, complex behaviour - that seemed very unlikely. And as matter isn't expected to do that, what we observe looks like curved space at work. But if matter is as described here, the velocity we see is not the whole story.

That works well, but there's still the football travelling through long grass. If matter really is like light, there's a need to explain why the football doesn't do what the torch beam does when it emerges from the tank of water, and speed back up again when it emerges on the other side.

The reason is that the football is being slowed by different forces arising at much larger scales. These forces work in an entirely different way - and many collisions, say, with the atoms in the blades of grass, will permanently reduce an object's momentum. So the large forces hide the small, subtle forces that are nevertheless quietly affecting things as well. The larger forces change the angle on the helical path, by slowing the object: it then doesn't change back. But the small forces leave the angle intact - the refractive index gets altered, which changes all three speeds on the helical path together.

So on Earth matter doesn't seem to behave like light. But out there in space, with just orbiting rocks and peace and quiet, all those larger-scale forces that mask the small-scale ones are removed. You just have matter, unimpeded by other matter. But there's a Planck scale refractive medium, which can get to matter at the scale at which it has a lightlike nature.

To test this, you need a refractive medium. But a large-scale one won't work, and we don't necessarily have anything approaching the technology to make a Planck scale one. But according to PSG we already have one here, the one that causes the Earth's gravity field. Later on I'll outline an experiment to test this idea, and find out if this medium is actually there.

40. Simple is hard, complicated is easier

This picture, with an emitted refractive medium, led to a gravity theory that mimics general relativity so closely that ten years of searching for differences turned up only a few. But it's far simpler, and it diverges in a few places.

The large-scale part of the picture appeared first. It was shaped by a lot of things, including mathematical clues from general relativity, and in particular one experimental result that started the theory. The mathematics and the conceptual picture shaped each other as they went along, each affecting the other. The result was a theory of gravity with a very detailed interpretation alongside it. The conceptual picture is far more detailed than in most gravity theories. That means it's going to be easy to check.

Many physicists are aware that general relativity has a mathematical 'simpler cousin', which is normally incomplete. It has usually seemed impossible for something like that actually to be right. What PSG does is to put changes into the background picture, which then allow it to work.

The result seems so simple compared to general relativity, it almost reminds you of something out of *The Flintstones*. They have technology that's oddly similar to ours, but in simpler form, and often using animals. PSG doesn't use animals, and it's actually far more sophisticated than it might seem. Although it's very simple mathematically, mathematical simplicity is a bonus. John Bell, when he found his breakthrough theorem about quantum mechanics in the 1960s, was accused of using high school mathematics to do it. It should have been a compliment. Often in physics, simple is hard to find, complicated is easier to find. But, as in the chapter on the good points Neil Turok has made about the present situation in physics, simple is what's needed.

41. Single points in space

Returning to general relativity's 'simpler cousin' theory, a lot of physicists are aware of it, but most see it as irrelevant. On the conceptual side, there might be a refractive medium - this is suggested because on the mathematical side, you look at what happens at single points in space. General relativity instead tends to look at the whole path through the field.

You find the result is the same either way, if you look at it in three large-scale dimensions (as we do when we observe things), and at single points in space.

At any point in space near a mass, light and matter are slowed by the same factor. This allows the mathematics to turn into a simpler version.

In fact, it's actually light, matter, and time, but I won't complicate things with that. In PSG these speeds are reduced together, and all by the same factor: they're multiplied by an expression from general relativity that gives various things about the field:

$$\sqrt{1 - (2GM/rc^2)} \quad . \tag{4}$$

In weak gravity such as on Earth, this gives a number like 0.9999999993, so they're only slowed very slightly. As you can see, the core premiss of PSG is simple. Light, matter and time are all slowed by the same factor at any point in space. Expression 4 has the radius 'r' in it (different from the 'r' term in equation 2), so it varies with distance from the mass, getting weaker further away. In general relativity it gives the time rate and the wavelength shift. In PSG it also gives the local speed at which space transmits waves.

The explanation that accompanies it is self-consistent, and checks well. The same expression sitting underneath the field can be arrived at in more ways than one. People don't usually look at what happens in general relativity at single points in three-dimensional space. But if you do look at that, you find that light, matter and time are all slowed in the same way.

One premiss of PSG is simply that this is not a side issue or a by-product. It's an aspect of general relativity, but it's usually not taken to be a central one. But PSG says it's the main thing about a gravity field.

General relativity has what behaves like a fourth large-scale dimension, but PSG says it doesn't exist. If you include it, and look at what happens in four dimensions, you get what are known as 'GR distances', which are longer than 3-space distances. GR distances are covered at normal speeds, without any slowing effect. It's as if the apparent slowing is just a trick of the fact that we see things, and measure them, in three dimensions, which hides some of the motion. But PSG says the slowing is real, and that with the right concepts in place, calculating using three dimensions, as in how we observe and measure things, shows what's really going on.

Although matter behaves in unexpected ways, the theory is straightforward. It might be surprising that it mimics the far more complicated equations of general relativity as well as it does, but it does.

General relativity has been tested for most of a century, and has passed the tests many times. So a theory that closely mimics it is needed. Anything that produces *very* similar results might be right. PSG is conceptually different, and far, far simpler. But if the observed results are truly the same, William of Ockham who (as in Book I) set out simplicity in all explanations as a principle, might have been drawn to this simpler explanation.

42. The time delay of light

In the 1960s people started bouncing radar signals off the inner planets. A radar signal is like a light beam, but of not the visible kind - the wavelength is longer. If a radar signal is sent off towards Mars or Mercury, it will bounce off the planet in many directions. But one of those directions is Earthwards, and a few minutes later a faint echo comes back, carrying information about the Earth-planet distance at that point on the orbit. If we know it's coming, we can find it. So for several decades, timing the delay (and dividing by two) has allowed us to build up increasingly accurate maps of the orbits.

But when radar ranging started, there was also a more interesting possibility. It was a way to test general relativity, which Einstein hadn't thought about. It involved sending a radar signal as near to the Sun as possible, almost grazing the edge but just missing it, when Mars was over at the opposite side of the Sun. It's the furthest Mars ever gets from the Earth, about 2.5 AU, because Mars' orbit is at 1.5 AU from the Sun, and the Earth's is at 1.

The signal would bounce off Mars and back to Earth along the same path, arriving back 40 minutes later as a faint echo. The experimenters might sit there drumming their fingers, knowing that on its journey the signal passes right near the edge of the Sun twice, and will bring back information about the huge gravity field there. The experiment also had to be done when Mars wasn't behind the Sun, to know the exact distance. It was then possible to get at the really interesting information - how much the signal gets delayed when passing through the strong part of the field.

In the wider universe the Sun is like a little burning grain of dust, but within the solar system it's enormous. Its mass is a few factors of ten up from those of the planets, with a gravity field large enough to reveal things that couldn't be measured on anything else nearby. In the general relativity picture, the Sun's field is like a huge crater in space itself, and sending a signal through that crater - or whatever else it was - was a fascinating chance to test out our ideas about gravity.

Experiments like this with radar had been done since the mid '60s, but they only started getting accurate enough to be interesting to theorists when the Viking landers got to Mars in 1978. Unlike with an orbiter, the distance from the Earth to a Mars lander was known very precisely. And they could boost the signal before sending it back.

The delay became known as the Shapiro time delay, after the man who did a lot of work on it. As the signal passed by the Sun, there were two separate effects that should delay it slightly, by an equal amount. The first is a delay due to time, the second is a delay due to space. The two delays are equal because of a broad symmetry between space and time in general relativity. The time part simply comes from gravitational time dilation - if the signal passes through the stronger part of the field where time is passing slower, it will be delayed on its way to Earth, by an Earth clock.

The space part of the delay can be seen in several ways. It can be a stretching of space in certain directions, but it's more helpful to see it as arising from the actual shape of the crater. The signal has further to travel, around the curve of the crater floor, than it has on the straight line distance, in three dimensions. The curved path might be called the 'scenic route', and whether the fourth large-scale dimension is real or imaginary, one can work out the extra distance the signal travels.

Taken that way, the extra distance to take the scenic route (the curved path minus the straight path) is 19 kilometers. The first point is: that's the kind of extra distance you might drive for a real scenic route. It doesn't fit our idea of the Sun's gravity field at all. We might visualise that kind of crater via how the trampoline stretches under the bowling ball. But if the curve is only 19 km longer than a straight line, it's not deep. It must be a very shallow crater, almost flat. The Sun is more than a million km across, and the strong part of the field, the crater surrounding it, is even larger.

The point is, gravity is a very minor effect, but it's enough to do what it does. A huge mass only affects space very slightly, in a range of theories. In general relativity the Sun's crater is shallow because a lot of matter only curves space a little. In PSG a large mass only makes the surrounding space vibrate a little. In both, matter only has a minor effect on space. So gravity equations often give a number just below 1, like 0.999999995. Without the mass there, that number would be 1, and space would be flat, or non-vibrating.

The expected duration of the delay to the signal was calculated, as predicted from general relativity. It's about 250 microseconds (millionths of a second), which is small, but very measurable. That's the time light takes to travel just under 19 kilometers, multiplied by four. The first factor of two arises because as well as the space component of the delay there's the time component, which is equal to it. The second factor of two is because the signal does all of this twice, on its way out and back.

So the experimenters waited for the chance to find out what would happen, wondering if they'd measure 250 microseconds. When the experiment was finally done, the result fitted the prediction perfectly. It was later confirmed more accurately by sending a signal near the Sun and to a space probe. This idea of timing light signals added an extra test of general relativity, and it passed. At the time, it seemed yet another good confirmation for the theory, arriving alongside a cluster of experiments with new technology that arrived in the '60s, half a century after general relativity began.

43. What led to PSG

And that result also led to PSG. Around 1998 or so, when the gravity part of the theory was just a sketchy group of ideas and questions, that result was one of several things that led to the picture of the secondary vibrations. The problem was, the space part of the delay wasn't explained in any way in the surrounding ideas, but it had been measured.

It seemed clear that space wasn't going to be curved at a large scale. There were reasons to think gravity is caused at a small scale - I had realised that inertia might arise at a small scale, and others in the '90s were thinking in the same direction. That could mean $F = ma$ also arises at a small scale, and if so, the equality between gravitational and inertial mass (which suggests a real force behind gravity), points in the same direction. So looking at this result, if there's no large-scale curving of space, instead there was a need for a *local* slowing effect, and one that could explain the measurement.

Whatever the unknown phenomenon was, it was possible to trace its effect, by translating between the delay that's measured along the whole path, and what happens at single points in space. The time part of the delay can be looked at either way, and one can translate between them. There's a delay to the signal along its whole path, making a quarter of the 250 microseconds. But we also know how time slows at a single point in space. It's just the gravitational time dilation factor, which as I said is

$$\sqrt{1 - (2GM/rc^2)} \quad . \tag{4}$$

So the total delay due to time is made up of many of those factors added up, for every point along the lightpath, with only the radius r varying. The space component turns out to be the same. It could be deduced that because the total delay due to space equals the total delay due to time, space must also slow radar signals somehow, at any given point, by the same expression. And it does - in general relativity as well, as observed in three dimensions.

And whatever the effect was, although at first glance it just seemed to slow light (ie. radar signals), it had to be able to slow matter as well. This could be worked out in several ways, some from the background theory I already had. So it was starting to look like something was reaching out from the mass, and affecting everything.

This led to the idea that some of the vibration that *is* the mass leaks out into space, or is transmitted to the surrounding space. It wasn't the only clue that led there, and other bits of the jigsaw were already in place around the hole. But when it was added in, the new chunk of jigsaw fitted with the other bits well, in quite a few places.

I've given an oversimplified version, but at the time the result from the Mars experiment helped with getting there. I walked around the edge of a large square field about a mile across, in Surrey one day in the late '90s, and came back with the ideas. By about '99 that part of the picture was in place, but the actual mechanism for gravity was five years away, and the mathematics that would later become the core of the theory was still more than ten years away.

But I found a way to show that the secondaries exist in 2007, and published a generalised equation for the geodetic effect, which allows one to see them at work. That's how it looks if you look at it that way, though it's not the only way to look at it. But to me at least, for rational reasons that were supported by a fair amount of cross-corroboration, by 2007 the idea that a mass emits small-scale waves had turned out to be right.

Part 9. Extra add-ons to Newton's work

44. The deflection of light

Gravity leads to side effects. Some are about light, others are about matter. There's a list of five main ones: for light there's the gravitational redshift, the deflection of light, and the time delay. For matter there's the perihelion shift and the geodetic effect. General relativity explains them all well - ironically, some would say it explains them better than the main effect, which is falling objects.

In the early 20th century, there were two effects that led to general relativity being widely accepted: the perihelion shift of Mercury, and the deflection of light by the Sun. The first was about matter, the second was about light. Both were measured differences from Newton's theory, but one was found before general relativity arrived, leading to a major puzzle in the 19th century, while the other was measured afterwards, and was used to confirm it.

The discrepancies were of very different sizes. With Mercury there was only a tiny difference to the Newtonian orbit, but the deflection of light led to a doubling of the original angle. Eddington's expedition to photograph a solar eclipse, and the nearby starlight that becomes visible, three years after the theory was published, loosely confirmed it. Since then, this doubling of the deflection angle has been measured much more accurately.

With the deflection of light, it's well known that a straightforward calculation shows the measured angle could be being caused by refraction. Using well understood mathematics, calculations have been done many times over the last century, by RM theorists and the RM-curious (including Eddington in 1920), ever since general relativity first appeared.

According to general relativity, both of these effects that helped to confirm it came from curvature. Both were effectively found within what for clarity I'll call the post-Newtonian adjustments, which were bits of mathematics added onto Newton's equations in the 20th century. They were largely about adding space curvature onto Newton's theory (Chapter 28). Curved space changed certain things, such as orbit paths, very slightly. Although they tended to be approximations for a more complicated theory, when the adjustments were added in, they fixed the discrepancies very neatly.

But instead of being about adding in a very slight curving of space, the post-Newtonian adjustments could be about adding in the refractive medium. It mimics curved space because the medium dissipates, which as I said before, behaves a bit like shading does in a drawing. It makes an object look curved into 3D, when it's really only in 2D. The equivalent of this with gravity is that instead of space being in 4D, it can be in 3D, with emission that fades as you move outwards. Either way, it only affects things a little.

In PSG the direct effect of the medium also needs adding in afterwards, and *separately*, just as curvature does. I'll show exactly why it needs to be added separately later (Part 18). This is important, because it actually interprets the entire structure of our present gravity physics, in which very small alterations need to be added separately. So it potentially sheds light on the relationship between Newton's work and Einstein's.

The two alternative add-ons are mathematically almost identical. In either, adding in what I mean by 'post-Newtonian adjustments' is exactly like adding in a slowing of light and matter at single points in space, in three dimensions (on Euclidian paths), by expression 4, as on page 69. Because of this, quite a list of effects could be caused either by curved space, or vibrating space.

45. The perihelion shift of Mercury

The extra perihelion shift of Mercury, which at the time seemed to provide confirmation for general relativity, was a divergence from Newtonian gravity near the Sun. It showed up in the strong part of the field. Mercury goes into strong gravity more than any other planet, so any anomalies might show up, including ones that wouldn't be noticeable further out.

Each time Mercury swung by the Sun, travelling on its really quite elongated elliptical orbit, something unexplained was happening. Some unknown effect was adding a small angle to the expected perihelion shift. It's amazing that without calculators this was a major mystery in 19th century astronomy, but it was. From many pen and paper calculations, they found that Mercury was getting a little extra 'tweak' each time it passed closest approach, swinging the position of its elliptical orbit around slightly each time. The orbit traces out a 'rosette' pattern over time, like a shape made by a spirograph.

The gravity from other planets passing nearby contributed, sending Mercury slightly off its basic path. But a small, steady shift of angle was unaccounted for: it was about 0.012 degrees, or 43 arcseconds. But that's per century - the effect is not a large one. The reason 19th century astronomers could see it at work is that angles get amplified over large distances.

According to general relativity, which provided an exact answer, this was due to space being steeply curved near the Sun. Mercury was going into a crater in space, curving around inside it, and coming out at a fractionally different angle from the one Newton would have expected. The mathematics fitted this explanation perfectly.

But what was happening could also be reduced to a straightforward slowing effect in three dimensions - of matter, not light. Very few expected it to be that, because matter doesn't do that on Earth. But whatever was causing the effect, what was *actually observed* was that each time Mercury went near to the Sun, on top of the acceleration from Newton's orbit, there was an extra slowing. And whatever that slowing down of the planet was, it was causing (or was closely related to) the shift to the deflection angle.

The slowing effect is graded in both theories, getting stronger as you go near the centre of the field. It might be caused by an unseen crater, or a refractive medium: as long as matter can behave like light at a small scale, it could be either.

And PSG produces orbits so similar to general relativity orbits, that looking at the data, it's hard to tell what we're seeing. This is particularly true because some of the parameters, such as the masses of the Sun and planets, are put into the ephemeris software without being certain of them. Some numbers are measured, but others we get by assuming general relativity is right, and then calculating them from the measured numbers on that basis. This is fine if general relativity is right, no problem. But if it's wrong, or partly wrong, as theories have turned out to be in the past, the real numbers might be slightly different. And these differences might allow any theory, if it closely mimics general relativity, to fit the data just as well - or perhaps better.

46. Five effects

The equivalence of these two alternative add-ons to Newton's theory, space curvature and refractive medium, means that either of them might explain a list of effects. I'll set the list out again - there are three light-related ones: the deflection of light, the time delay, and the gravitational redshift. And with PSG, in which matter can respond to a refractive medium as light does, two matter-related effects now become explainable as well: the perihelion shift, and the geodetic effect.

PSG covers far too many areas for it to be complete. If you get to a theory of everything, you might get the theory, but you won't get everything. And to make my excuses for it being incomplete in some places, I worked very hard on another area of it, DQM, dimensional quantum mechanics, over the last few years - while the first paper and the first book were being written, trying to finish them while the documentary on them was being made (and to add some points I'd found as a result of the conversations for the documentary.) With these effects, there are places where I can see exactly what PSG does, and have tested it out with calculations and found that it works perfectly, but without always getting to generalised equations. But in other areas, I did get to generalised equations.

And there's a difference. At one point a few years later on I tried to publish a calculation showing that the geodetic effect could be caused by a refractive medium, and no journal would take it. It was turned down again and again. But a few months later I got to a generalised equation showing exactly the same thing, and it quickly got through peer review.

Anyway, clear mathematics came out of the theory in many other areas. And generally, across several effects, wherever general relativity adds curvature onto Newton's theory, PSG adds the refractive medium instead.

I'll repeat the underlying similarity: in both theories the add-on can be taken as a slowing of light and matter at single points in space, on Euclidian paths, by expression 4. The two theories are mathematically equivalent in exactly this way in many places, which is why the well known effects of gravity could be either.

Part 10. The actual force

47. The inverse square law

The inverse square law for the force of gravity is one of the most basic laws we have. It has no clear explanation - either from within general relativity, or from outside it. As often happens, general relativity can only show exactly how it comes out of the mathematics, and it certainly does.

John Baez, a well known relativist who incidentally is also Joan Baez's cousin, set out a proof that in weak gravity Einstein's equations reduce to Newton's, with the inverse square law. Mathematics of this kind shows part of the story - Newton's theory emerges from Einstein's. But it would: the way that space curves was set to reflect what we already knew about gravity, or rather, was found to work when compared with what we already knew.

And the question remains. For whatever reason, the force of gravity weakens with the square of the distance. Take an asteroid - a chunk of rock, say about the size of a bus station. Whatever the strength of its gravity is a mile away, at two miles it's four times weaker, at three miles it's nine times weaker, and so on. The force fades rapidly with distance.

Inverse square laws can be found elsewhere: in electromagnetism, and more generally when waves are radiated evenly outwards from a source. We have a good understanding of that principle, and it explains most inverse square patterns we know. Some think it also fits gravity, but many think not.

When light moves away from a light bulb, it only fades because it gets spread thinner as it moves outwards. As it travels away in all directions, the overall intensity stays the same, but spread across a sphere of light that expands as it moves away. The sphere's radius increases evenly over time, but its surface area enlarges faster and faster.

So if you look at the light from a particular place on this sphere, and people often do, the amount of light you get is the total, *divided by* the surface area of the sphere at that distance. That leads to an inverse square law, because the surface area is $4\pi r^2$, so the strength of the light is proportional to $1/4\pi r^2$, so also to $1/r^2$.

With gravity, the question is, what causes the inverse square law in the force and acceleration? You might ask if in PSG the secondary vibrations behave as with the light bulb, going by the 'surface area principle'. It doesn't seem so. What we know about the secondaries so far is how they slow things down at any point in the field, although I sometimes use the word 'density' in a loose way. They may be analogous in this area to other things we know of - they certainly are in some areas. Their slowing effect, as in the subtracted speed, is either way almost exactly proportional to GM/rc^2, so to M/r. And overall, their slowing effect is just expression 4.

So their slowing effect doesn't go by an inverse square law, and it seems the 'density' doesn't either. If the secondaries behave as refractive materials on Earth do, the density to refractive index relationship will be roughly linear, as that's what happens at lower densities. And it seems their density is *very* low compared with materials we know.

In the context of PSG, this could be part of how the nature of gravity stayed hidden for so long. Anyone investigating the inverse square law, if they think of an emitted RM, and look for the 'surface area principle', will get to a dead end. Perhaps that made people not look any further. According to PSG, the actual reason for the inverse square law is more indirect, and can't be found without taking one or two other steps first. But when you get there, it works very well as an explanation.

48. Why is the Earth stronger than the Sun?

The PSG picture suggests a reason for the inverse square law. To get a sense of how this works, it's worth looking at the bizarre question of why we don't fall towards the Sun. At the surface of the Earth where we live, the two main gravity fields affecting us are the Earth and the Sun. They combine, and in many ways of comparing them, the Sun's field is the larger one.

The Earth is nearer us than the Sun, it's right here. But even allowing for that, at the surface of the Earth most of the numbers for the Sun's field are higher - in just about any theory of gravity.

In Newton's theory, a relevant number is GM/r, and at the Earth's surface, the Sun's gravitational potential is 14 times larger than the Earth's. In general relativity, the Sun's 'crater' in space is effectively deeper, and the spacetime curvature affects the time rate more. In PSG, the Sun's mass vibrates space more, sending stronger vibrations to the surface of the Earth than the Earth itself does, and slowing wave motion more than the Earth does.

So it's reasonable enough to ask why we fall towards the Earth, and not the Sun. It's because the strength of gravity isn't about these numbers, it's about how fast they're changing.

The trampoline analogy from general relativity, though flawed if one tries to get it to explain gravity, is still good for picturing gravity. Imagine a ping pong ball placed on the trampoline. It'll be pulled more strongly into a small crater with steep sides, than into a large crater with not so steep sides. Putting it that way, it immediately becomes understandable, intuitively at least, why we fall towards the Earth. The Sun has an enormous crater, as gravity craters go in the solar system. But although it's comparatively deep where we are, at 1 AU its sides have a flat slope.

By contrast, the Earth's own crater is like a little pockmark inside this large flat crater from the Sun, but where we live its sides are steep. At the surface of the Earth, which is a point quite near the centre of the field, the numbers for the Earth's gravity field are changing rapidly.

The slope of the sides of one of these craters, or indentations in space, is like the slope on a graph. How steep it is at any point is about a *rate of change* for something that has somehow altered the state of space out there. Both general relativity and PSG give hints that the strength of the force of gravity is related to this gradient somehow.

In general relativity, if you look at the idea that the force of gravity and the slope of space curvature are linked, this might be about how the contours of space affect a moving object: there's no real pull towards the centre, but the paths objects take suggest one. But for an object stationary in relation to the central mass, there's a problem explaining the force that pulls at it. If one relates *that* force to the local slope in general relativity, the slope starts to look too much like a real slope, as if the analogy is being taken more literally than anyone wants to take it. And no explanation appears.

In PSG there's a path that leads towards an explanation. As in most gravity theories, PSG has different sets of numbers relating to the field. Approaching the mass, some get lower, some get higher. And there's one set of numbers that changes so its slope gets steeper as the force of gravity gets stronger. It comes into general relativity as well, and can mean various different things in both theories. And it has an inverse square pattern hiding in it.

This set of the numbers is the rate of change, with radius, of several things. There's the time rate, the gravitational redshift, the energy, the effects of frequency differences on other quantities (the time rate and gravitational redshift can be alternative versions of the same thing). These all might be possible culprits for the inverse square law. They come into both theories, and the mathematics is the same in both theories, for all of them.

Whatever this set of numbers means (expression 4), it has an inverse square pattern in its rate of change. In PSG, as well as those meanings for it, there's something else, which leads to an explanation.

It's the rate of change of the transmission speed of space, also inversely the refractive index of space. Now 'refractive index' is a human made up system, but in PSG it's related to the vibrations that fill space, and how much they slow anything down. That ability to slow things down increases nearer the mass, and the rate of change of this ability makes a slope that's very steep at the Earth's surface - it's far steeper than the Sun's slope there.

And the transmission speed of space starts to fit the bill. It's a quantity that comes into Snell's law, and its rate of change is more or less proportional to $1/r^2$. To many decimal places, it's proportional to the force of gravity.

So there's an inverse square relationship in this gradient in the numbers, and another unexplained one in the force of gravity. The need is to link them up, and PSG has a way to link the inverse square pattern in these numbers with the other one - the one in the actual force.

49. The force that pulls the apple

Let's go back to the apple orchard, and look right into the apple in the tree, just before it falls. Its matter is still on horizontal circular paths, but a force is trying to pull it downwards, onto helical paths. Refraction is trying to shift its matter to a slight angle, but the apple is held where it is, at a large scale by the stem, and at a smaller scale by the electromagnetic forces that hold bits of trees together. These forces are much stronger than gravity, and tend to hold everything together.

So the waves keep circling. They can't just go spiralling downwards, as would naturally happen. But they try to, because the space behind the apple has a kind of slope to it.

Now at the height of the apple in the tree, the steepness of the slope there (the rate of change of the way space is vibrating), affects the angle through which refraction is trying to pull the apple's waves. And that will affect how steeply they're trying to dive into the refractive medium. And that affects the strength of the force pulling on the apple. The faster the refractive index of space is changing there, the more space is trying to pull the apple's matter through a steeper angle change, to a faster downward speed, which means with a stronger force.

It's like the laser beam in the tank. The way the refractive index of the sugar solution is changing will affect the way the angle of the laser beam changes. So although it's not quite so simple, it's significant that the rate of change to the space behind the apple varies in proportion to $1/r^2$. And it becomes clear why gravity follows an inverse square law.

Ibn Sahl's law of refraction applies at all scales, and for waves of all sizes. As described earlier, any refracted wave has a speed difference across its width, between one side, or edge, of the wave and the other. The greater the ratio between the speeds, the stronger the force that pulls the wave sideways. As in the law of refraction, there's a general aspect to this, which affects a wide range of different situations in the same way. And according to PSG, it leads to the force of gravity.

50. A key, and why gravity is like velcro

Whether or not this is the true explanation for the force of gravity, we know the strength of the force is affected by something else as well. If the apple's mass was three times larger, the force pulling it would be three times larger. Unlike the acceleration, the force, GMm/r^2, has m in it, because the force is also shaped by the size of the passive mass. It gets multiplied in, and in this case that means the apple.

The value of what's called the 'small mass', is a measure, loosely speaking, of how much matter is in the object. In general relativity it's not easy to explain why it affects the force, particularly when the apple is still in the tree. In PSG mass directly affects the force of gravity, and is proportional to it, because in this situation the value of the mass is about how many of these little helixes are combining their efforts, trying to get refracted, and so all pulling towards the central mass together.

So in the picture that emerges, the particles inside an object are not quite as passive as the phrase 'passive gravitational mass' suggests. Each of them is directly contributing when a large-scale object is pulled towards a mass. Each has a very small effect, but there are trillions of them, and it adds up. One can imagine the apple, or try to imagine it, as an enormous number of these small helixes clustered together, all 'spiralling' away, and all helping to drag it through the surrounding gravity field.

So gravity is like velcro, with its many little hooks on one surface, and loops on the other. Both sides that stick together, or in this case pull together, play a part in the process. That's why gravity has been so hard to crack into and understand, according to this picture of it. You need both parts of the key for it to turn, and RM gravity was only one part. Without the nature of matter as rotating light, which is the other part, you can't get to the small-scale picture, and although the key looks promising, it doesn't work.

But when you have both parts, it opens the door, including mathematically. As always, the key itself is conceptual, not mathematical - that's physics for you. And where general relativity's twin turns out to be hiding, involving a real force, and not a pseudoforce, was a place that they couldn't get to in the early 20th century. It's not just the Planck scale, they had some idea of that. Theodor Kaluza sent his mathematics for a 5th dimension to Einstein in 1919, who later sent the paper in on his behalf. By 1926 Klein had suggested small circles, and the idea was born. But what they couldn't get to for another sixty years, was matter being small circular loops, as in string theory.

But by 1984 that idea was on the map, and according to this view of things, that part of the key was then findable, by taking one more step - realising that these loops are not separate from the dimensions, but places where the dimensions have circular vibrations going around them.

In PSG, all matter is waves in the fabric of Klein's circular dimension, rotating around it. If so, matter is very like spinning light. And if so, refraction - which although it bends light, had never been seen as a force - can become a force. You can then reach an explanation for gravity involving a force, but one that had not been taken to be a force in the past. But in a different context, with some essential concepts shifted around, it turns out to be one - explaining several things at once, including why it wasn't found before.

51. Getting to the common root

Newton assumed that gravity affects an object via its mass, and it does. But at the time, no-one knew about the mass-energy equivalence, $E = mc^2$. Three centuries later, we know mass and energy are directly interchangeable: mass has energy, energy has mass. So gravity might affect matter via both. Most of the equations are in the mathematics section, but I'll briefly mention one (from Chapter 69) here.

Over the last few chapters, I've said the ultimate cause of the force of gravity is *the local rate of change of the transmission speed of space*. That, according to PSG, is what creates and quantifies the force. So it's significant that this rate of change produces something very familiar, if you just multiply it by the apple's internal energy: its mc^2. That simple setup - two numbers multiplied together - gives Newton's formula for the force of gravity.

Standing further back to sum up, in the picture I've described, gravity works on matter at a fundamental level. This common root is where mass, inertia and the basic aspects of matter appear, and all matter comes together under one umbrella. Without something unifying all matter, it wouldn't behave as it does. This common root seems to be near the Planck scale. String theory has already cracked into that scale mathematically - with an incomplete picture accompanying it. So we may need to fill in more of the picture before we can make real progress. People very often try to push the mathematics forward before the picture is well in place, but it usually doesn't work.

String theory looked hopeful for years, but as progress continued, it divided into a large number of theories, 10^{500} or more, and there was no way to tell which was the right one. Even worse, people had confidently predicted that the Large Hadron Collider, when finally built, would confirm supersymmetry, as string theory *requires*. But it found nothing like that, and this weakened the whole edifice of string theory. In 2020 a NASA experiment, searching for axion-like particles to test string theory, also found nothing. Many physicists now think string theory is stuck, or worse.

So mathematics is not getting us there - we need a *conceptual breakthrough*. Nowadays, many physicists are working on new mathematics for the Planck scale. They're trying things out like alternative rules for the geometry there, but this is still a totally mathematical approach, and it's still like looking for a needle in a haystack. The theory in this book has mathematics there working exactly as it does elsewhere, but instead there are conceptual differences, which involve the dimensions.

Meanwhile, the Planck scale is quickly becoming a new layer, added to what we call science. There's a series of layers: going up the size scale, each layer emerges from the one below it. Going down the scale, each layer seems to reduce to the one below it. It's thought that biology reduces to chemistry, chemistry to particle physics. And it seems that particle physics reduces to Planck scale physics.

So filling in our picture of the Planck scale is important. It's the deepest layer - for now at least. This new view of gravity and quantum mechanics, PSG and DQM, describes the Planck scale, or an equivalent scale near it. It has some unexpected twists, and hopefully shows more about what the world down there is really like. It also explains something that in standard physics is still a major mystery: how what happens there makes the world we know emerge at all - as we know it does - further up the scale.

Part 11. Gravitational waves

52. Ripples in something

Nowadays people try to avoid the cumbersome old word 'gravitational'. We say things like 'gravity field' instead of gravitational field, and have been able to simplify it in that instance - it's clear enough what's meant, so the phrase is quite widely used.

People also say 'gravity waves', but it's not so good. Gravitational waves are disturbances in a gravitational field, but gravity waves, in the strict use of the term, is a general wide ranging type of wave, which includes ordinary ocean waves on Earth.

I doubt if there have been too many misunderstandings between physicists, or hilarious, farcical conversations at cross-purposes. But once in an article on waves deep within stars I've seen the term looking potentially ambiguous to some readers. Anyway, I've mentioned it now because I'll use the informal version here (incorrect unless mentioned), which is easier. So 'gravity waves', in what follows, means gravitational waves.

Gravity waves are emitted if a mass changes its shape. In general relativity, the surrounding field is like a curved indentation in the fabric of space. If the mass changes shape, the field does the same, and the new pattern spreads outwards, replacing the previous pattern at the speed of light. So the fabric of space adjusts itself via a 'wave of curvature'.

The trampoline analogy helps. Say something happens with the bowling ball - waves will travel outwards at the transmission speed for waves across the rubber sheet, carrying information on what's happening at the centre. One side issue - though it's rarely stated - is that because in the real world these waves travel at lightspeed, it's clear that according to general relativity, the transmission speed for waves in the fabric of space is c, as it is in PSG.

In the universe, when a mass changes its shape it has probably exploded, or collided with something. If that kind of thing happens, the mass will probably go through a series of rapid changes to its shape, and that leads to a series of ripples in the field.

In PSG the picture is different. Before the mass does anything, it already has emission pouring out of it. The mass is like a sparkler, or a spherical fountain.

It is emitting very small vibrations, and they travel outwards, fading slowly as they go. So far the mass hasn't even changed its shape yet.

Gravity waves are not this Planck scale radiation, they're enormous patterns in it at larger scales. If the mass changes its shape, corresponding patterns in the radiation moving away from it appear. They move outwards at the local speed of light, because the fountain itself, on which the patterns are written, is moving outwards at that speed.

So gravity waves are large waves carved out of a sea of small waves. And the small waves are waves emitted by other vibrations - it's clear enough what our world is made out of in this picture of it.

As always, general relativity is constantly right about the mathematics, and usually about the observables as well. Where PSG says there are differences, they're sometimes just in the picture, and not measurable. But occasionally you get differences that can be observed. To me these are rare and exciting: I've hunted them for years, rather in the way that rare species are hunted by botanists and zoologists. I've probably missed a few, maybe many - it's hard to think of everything. But there's certainly enough for experiments already, and those will hopefully show which of these two pictures takes us nearer to whatever it is that - as they say - is actually out there.

53. Ways to find them

There are two main ways to detect gravity waves, as they're expected to be in standard physics. The more direct method is about looking out for changes to distances in the space we live in, by monitoring the distance between two objects.

If space itself expands and contracts, the idea is that as a gravity wave passes by, it should briefly compress the length of everything in one direction, and expand everything in a direction at right angles. And nowadays 'everything' includes objects like metal tunnels a few miles long, with lasers monitoring their exact length. This change to the size of our world in certain directions would be quick and very small, but it's measurable.

The more direct approach came first, and people originally used chunks of metal. Starting in the '60s, they monitored them to see if they changed size when a wave passed by, but it was too early to find anything. But then a less direct method appeared in the '70s, and showed without doubt that gravity waves exist. It was about measuring the energy loss of a pair of stars orbiting each other closely, to see if they're giving off gravity waves.

Two American astronomers, Russell Hulse and Joseph Taylor, discovered a spinning pair of pulsars in 1974. They spent the next few years analysing the rapid pattern of 'bleeps' or flashes coming from them. The stars were moving towards each other, each also slowing down on its own axis. So the flashes were slowing down, giving a very accurate account of what the system was doing. They were able to work out the energy loss from that, and it neatly matched the energy the system would be losing if it was giving off gravity waves.

A graph online (in the reference section) shows how it was then monitored from 1975 to 2005, the steady speed shift beautifully matching theory all the way along the curve. And since that first binary pulsar, eight more have been found. Whatever they actually *are*, gravity waves exist alright.

That work led to a Nobel prize. It also led, in the 1990s, to people starting to build sophisticated interferometers all over the world. LIGO is the largest of them. The arms are four kilometers long, and look like giant pipelines. From the air it's a huge L-shape, spread across the ground in Louisiana. There are two similar ones in Washington State, each with a pair of tunnels at right angles. In each, lasers are sent down the two tunnels, reflected off mirrors at a very exact distance away, and recombined back in the middle. The system is enormously sensitive to changes to the lengths of the lightpaths.

These interferometers often work together. They're looking for waves from distant events like black holes colliding, which should set off interferometers across the planet when they pass through the Earth. So LIGO also works with the VIRGO detector in Italy and GEO in Germany.

There are other indirect methods, such as attempts to detect gravity waves from the big bang, by analysing the polarisation of the background radiation. These seemed to have been successful in 2014, but the BICEP2 result turned out to be just the effect of dust within our galaxy.

There are other more direct methods. The International Pulsar Timing Array involves a group of astronomers working together, timing the flashes of light from pulsars. If a gravity wave passes by, they should find a pattern. Pulsar timing is different from the spindown rate method, although both involve pulsars. It's more like LIGO: the light from the pulsars is used like the lasers in a large interferometer, stretched out across space.

There are also two space based interferometers on the way, but LIGO is the best source of data at present. It has been getting signals from distant events steadily now - but we've only just started gathering the truly interesting data, and we've already found several anomalies.

54. The first signal from LIGO

In September 2015, LIGO finally got a good detection of gravity waves. There are sometimes false alarms, but this looked like the real thing. It was a quick series of ripples, $1/5^{th}$ of a second long, which hit two detectors at opposite sides of America. The time lag between them was 7 milliseconds, within the window for a wave travelling at lightspeed reaching Earth at a random angle, and sweeping across the two stations.

The wave could be coming from any direction, so the shortest possible time lag is zero. That's for if its leading edge happens to be parallel with the line between Washington and Louisiana - it reaches both at the same instant. The other extreme is at 90 degrees, and the wave travels along the line between them. It goes through the Earth, rather than following its circumference, but both routes are about 3000 km. The maximum delay is that distance over c in km, 3000/300,000, 0.01 seconds, 10 milliseconds. So the range of durations for the time lag is 0 to 10 ms, and 7 ms means the wave was at one of the angles between those extremes.

The signal turned out to be real, and was later confirmed by other signals. In the context of PSG, there were some interesting possibilities. Chapter 3 had an outline of the PSG version of things, which should hopefully make better sense now. In PSG what ripples is not distances, but the density of a medium, so what also ripples is the speeds at which light travels through it. In ordinary physics on Earth, this happens all the time.

Imagine a gust of wind blowing across a car's headlights. Before it arrives, the air in front of the car is still, but it slows the light beams down anyway, in the way that air slows light. Then the gust arrives. Say it has a higher density of air molecules than the surrounding air, as it might have. If so, the beams will be slowed a little more as the gust goes by, because the light briefly travels through air with a higher refractive index. So the beams travel slower, then faster, maybe several times, and the speed of light will ripple in front of the car.

Compare that scene, involving well understood everyday physics, to the idea that space stretches and contracts. It might, but there are no other examples in which we know for sure it does. There are other areas where it seems to, but there's always ambiguity. (People thought, after the geodetic effect was measured, that it was proof of curved space, but it later turned out that a refractive medium could also be the cause.) Still, this much is clear: whatever LIGO measured, it was probably one of two things. It was either variations to distance or speed - there's not much else it could be.

It certainly could be a speed variation. If what passed through the Earth that day was a ripple in a refractive medium, just as the gust of wind was, or as variations in the ocean density are, then the laser beams would be slowed in their tunnels, then speeded up, a few times. It creates a miniscule effect, but it would neatly mimic general relativity - it behaves exactly as if there were a series of changes to the length of one of LIGO's arms.

55. A major anomaly

So can we tell which theory is right? If one of them actually is right, or nearer the truth than the other, it may show up. One way is if we know something about the source from ordinary astronomy as well.

The signal that was detected in 2015 came from an unknown source, thought to be more than a billion light years away. It had the hallmarks of two black holes colliding, after spinning around each other rapidly and moving inwards. Somewhere out there, we think that happened - and it almost certainly did. But everything we know about the source was reached by assuming general relativity is correct, then interpreting the signal on that basis. This includes the masses, the distance to the event, when it happened, and so on.

All those numbers are in one sense unknown, although they're quite likely to be right. But general relativity has only been confirmed over short distances. To many physicists (who have an open mind about the unresolved question of dark matter), it has failed over long distances so far. And the distance was enormous - more than a billion light years.

And the numbers would look more certain if there had been fewer anomalies surrounding the source. Eventually there were three. It's the first detection of merging black holes, and the masses of the two objects, around 30 solar masses each, was a surprise. Stellar back holes are not thought to get that large - it later became clear that we had no mechanism at all for a binary that large to form in the first place. According to current physics, given their size, they'd have merged already, before even becoming black holes.

Nevertheless, some ideas about the source work fine if you assume general relativity is correct. But if it wasn't, you simply wouldn't know. So in certain aspects of the measurement, it's not appropriate to assume it. On the other hand, it's utterly reasonable to calculate the general relativity picture, and set out the numbers - as long as you don't then turn around and use that, by implication, to *confirm* general relativity, as some articles did at the time.

In fact, the mathematics and the implied numbers in my picture may well be almost identical, and as a later chapter shows, it seems that they are. But it's worth pointing out that since the 1970s binary pulsars have given far better evidence for gravity waves, *because we know about the source*. We watch it over decades, and we can check one thing against another.

So what's needed is a gravitational wave detection, but from a source that other kinds of astronomers can see as well. And although it may be too early to say, gamma ray astronomers did find an electromagnetic counterpart to the first signal from LIGO. The orbiting Fermi telescope, which covers a wide area, was looking in the right direction that day. Although other instruments didn't see it, there was a gamma ray burst, coming from the same direction in the sky, also on the 14th of September 2015.

The photons arrived at almost exactly the same time, reaching the Earth only 0.4 seconds after the gravity waves. Both travel at the same speed, and both were coming from the same direction, to some level of approximation. The two measurements look like they're the same event, and it was calculated that the chances of them *not* being from the same event was only 0.22%. The puzzle is that merging black holes, in general relativity, simply can't give off gamma-rays.

But merging neutron stars do. The astronomers using the Fermi telescope say in their paper that the short gamma ray burst was very loosely similar to other GRBs they've been looking at. But oddly, the brightness of the event, in those terms, did not come near to matching the distance implied by the LIGO measurement. It was very dim, and might have been, for instance, a pair of neutron stars merging, further away than the distance estimated via LIGO. But there was reason to think this was a pair of black holes.

A gamma ray burst from a merging pair of black holes probably means one thing: matter is present. Matter, when it gets smashed up in a very powerful collision, gives off gamma rays. But general relativity says that all the nearby matter should unavoidably have been swept up before that, and would have crossed the event horizon way before the collision.

So this was a major anomaly in certain ways, but most of the general public who read the coverage about the LIGO measurement knew nothing about it. They got the impression that everything was neatly tied up. But meanwhile, in the science media, and in the physics and astronomy literature, attempts were being made to explain it. Wild ideas were then talked about as if they were reasonable, such as both black holes having been inside an enormous star at the time. Many physicists think this is far-fetched, to put it mildly. But some articles in the science media talked as if this was the sensible, likely explanation. As often happens, the extent of the problem is illustrated by the desperate remedies that are put forward to explain it.

When the news of the anomaly came through, I looked at some possibilities. But after a lot of backwards and forwards, I ended up not believing that the problems about the source were caused by the numbers being wrong, even though further problems about it appeared later. Mathematical differences between the theories would have been exciting, as astronomers would keep on finding them. But there was another explanation from PSG that looked far better. I had believed it for some years, since around 2012. And it was also the kind of thing that astronomers would keep on finding.

But the anomaly nevertheless highlighted this interesting point: we've seen gravity waves at both ends of the (metaphorical) tunnel, but so far we've not tested the connection between them. Since the 1970s, gravity waves have been seen getting emitted from a mass, and the emitted energy was exactly as expected. But we never followed the waves, and we don't know for sure what size waves that energy created.

And now, here at the other end of the tunnel, we see gravity waves detected far from a mass, and by working backwards from their size, we can calculate what the energy should have been - and so also the mass, the distance, and so on. But again, the link between emitted energy and wave size hasn't been confirmed. So we've seen each end of the tunnel, but not the middle. You might say our only glimpse of one end seen from the other, in forty years, is that unexplained, anomalous gamma ray burst. It's worth mentioning that, and worth bearing in mind. But for reasons I'll give, my view is that general relativity will be right about the middle of the tunnel as well.

And anyway, this goes the wrong way. If the gamma ray burst was indicative of a difference of that kind, and was from, say, a distant neutron star merger, it would mean that small masses give off larger waves than thought. And yet the estimate for the gravity wave background of the universe was revised *downwards* a few years ago, due to not finding it in the range where it was thought to be. So this kind of idea probably doesn't work - I'm showing it to you in case the surrounding questions are of interest.

The waveform of the detected gravity wave was like merging black holes, not neutron stars. Anyway, the non-scientific press didn't carry the story, and the public were left thinking everything was neatly explained. For the time being, the anomaly leaves a hazy surrounding ring of question marks, and the thing is, they're in an area where many felt the question marks had finally been removed.

56. Looking for differences

It's exciting to compare the two theories closely, looking for differences, and in particular the rare and special differences that are observable. It is to me, but the trouble is, you keep finding how ridiculously similar the two pictures are. Of four possible differences about gravity waves, two were more or less ruled out, a third started to look increasingly unlikely, and the fourth, though it exists, seems to be unobservable.

In the general relativity picture, when a wave passes by and stretches space, the strain term h is the difference in distance between any two objects, as a fraction of their normal distance. So the factor difference varies from $1 + h$ to $1 - h$, as space gets compressed and expanded by the wave. But the number $1 + h$ includes two components, one for space and one for time. In general relativity, both are affected. There are different mixings between the two in different coordinate systems, which are different ways of looking at it. But the two combined always gives $1 + h$.

When interferometry was first proposed as a method for detecting gravity waves, people had to study general relativity very carefully, to make sure the space and time components didn't cancel, making the waves unobservable. Luckily, although the wave alternates between compressing space and time and stretching them, it turned out that it does this in a useful way. When it stretches space, so that the laser takes longer on its journey, it also stretches time, so that it takes even longer (and the equivalent happens when space is compressed). So space and time don't cancel, they double up, and there's a measurable effect.

In PSG, the changes to time are just the same as the standard ones. But with space, whenever in general relativity distances get longer, in PSG instead the speed of light gets slower (either way, the light takes longer to return). And whenever in general relativity distances get shorter, in PSG the speed of light gets faster (either way, the light returns more quickly). Now one difference between the pictures is that in PSG, h is composed of two *equal* components. The contributions from space and time can vary in general relativity, but not in PSG. I wish I could say this is measurable, but it seems not to be. Either way, you always just measure $1 +/- h$.

There are other possible differences. Gravity waves in general relativity are carried at c by a medium, but the density variations in PSG are in an emitted medium, made of waves, that's travelling at c anyway, through an underlying medium. So gravity waves are large waves in a moving sea of smaller waves, and they're waves of variation in its refractive index. I've believed that since 2004, and published it in a peer reviewed journal in 2008.

This means that gravity waves in PSG are, to some extent, emergent. They're not transmitted directly by an elastic medium, only their component waves are, at a very small scale. So there are some potentially quite drastic physical differences, which might lead to measurable differences - to how the waves behave in each theory.

57. A further mystery

Of the two avenues about the anomalies that LIGO had started to find, my preferred avenue was definitely one that I'll get to, which is about the nature of black holes, near the event horizon. Mathematical differences relating to gravitational waves seemed more of a long shot.

But there is one intriguing point that came later, which made the avenue of mathematical differences a little more likely again. Calculations showed that the two black holes that were meant to have collided were so large that they couldn't originally have been a binary star system at all. A pair of stars that large should have merged into each other before even becoming black holes. This means that although the impression for many was that everything had been explained and understood, the first detection by LIGO was absolutely beset with anomalies.

And later on I'll mention another anomaly, found within the LIGO signal itself later that year, 2016. It's a sort of echo that comes after the main signal, and some say that if it's confirmed, it would mean general relativity needs to be replaced. But the lack of coverage on that rather dramatic finding was not a conspiracy! No-one knew about it until two years later.

Incidentally, to me incomplete, over-compressed coverage of science issues is understandable - many of the details are too technical to go into, and they take up space. But news sources do tend to emphasise the triumphs, which is sometimes because that makes the story in the first place. The triumphs are real, and very much worth covering. But when readers want to know more, and find out about the actual struggle to understand the universe, with all the accompanying unknowns, difficulties, unexplained clues, and loose ends not yet tied up, that's also a good story.

The trouble with oversimplified science coverage - in non-scientific media - is that it can add up over a number of articles. One article might be only slightly misleading, but if things tend to be oversimplified in the same way, perhaps exaggerating confidence in standard theory, the overall impression can be more misleading than a single article.

Going back to the real story: in 2016, attempts to understand the source of the signal, and how it formed, quickly went to centre stage in astrophysics. And two competing theories started fighting it out, to explain how the black holes formed in the first place. The long-standing 'common envelope model' could explain the 10 solar mass black holes of LIGO's second detection, but had trouble dealing with the 30 solar mass ones of LIGO's first detection. A new theory called the 'chemically homogenous model' was exactly the other way round - it could explain the larger black holes, but not the smaller ones. Weaknesses were pointed out in the new idea, then ideas involving magnetic fields were put forward to get out of the problem.

The fact is, there are so many conceptual variables you can play with them indefinitely, but you might never get to what's actually going on. Perhaps the double black hole wasn't a binary system beforehand at all - some even think what happened is that the two of them met by chance, they were attracted to each other, nature took its course.

The difficulty explaining the source was interesting. At first it suggested real divergences from standard theory - in the numbers. But that's just a remote possibility, as I see it at present. Instead, the problems explaining the source may arise because black holes are simply different from general relativity's picture of them. This is borne out by the recent discovery of a 68 solar mass black hole. It's nearly three times the size that stellar black holes are thought to reach, and standard physics struggles to deal with it.

In the section on black holes, we'll look at one particular difference between the theories near the event horizon radius, which would explain not only the problem I've just outlined, but also a list of similar anomalies.

Near a black hole our theories splay out, and small differences get amplified. And the PSG picture has a good explanation for how matter survives near the event horizon radius, allowing it to get smashed into a flash of gamma rays at the point of collision, sending out a signal to electromagnetic astronomers (in this instance to Valerie Connaughton, the astronomer in charge of the telescope, who picked out the flash of light) that was - according to standard physics - inconsistent with general relativity.

58. Another optical counterpart

And then, in the summer of 2017, a pair of neutron stars merging was seen. It was an exciting time, as the event was detected both via gravity waves and also via an electromagnetic counterpart, a gamma ray burst. It's early days, but the mystery was that the GRB was far, far fainter than equivalent GRBs.

Again, news sources didn't mention it, and underlined the triumph of the measurement, and a triumph it certainly was. But meanwhile, astronomers and physicists were grappling with the extreme faintness of the GRB.

So at that point, two anomalies had potentially suggested that distances to the events are larger than thought, or the events are smaller than thought, for some reason. Two GRBs had been *way* too faint, and both of them were associated with a LIGO detection of gravity waves. If the distances are wrong, and/or if the black hole masses are wrong, that might perhaps remove the problem explaining the source of the first detection. This at first seemed a promising avenue.

But the two detections were different, in several ways. And to me, with the neutron star merger, despite this odd thing about the source, it seemed that the host galaxy had been identified, and that means we probably know the distance. If so, the faintness isn't due to the event being further away than thought. Perhaps it was a selection effect, where we focus on a faint flash of light that would otherwise not be noticed, because we also detect the gravity waves.

So taking all the clues together, which is what you've got to do, it looks like the events detected via gravity waves, though hard to explain, really were at the estimated sizes and distances. But time will tell, and LIGO has now had another upgrade, so there's better data on the way.

Part 12. Mathematics of gravity

59. Mathematical simplicity

Having got to the mathematics section, which I'm sure some will have been wanting to get to, I'll call general relativity GR, and Planck scale gravity PSG. This section is about the places where PSG is different from standard theory. It's about zooming in on those rare bits of the picture that I've been combing through it to find - which lead to new mathematics. It also has a near-proof, or what's known as 'smoking gun evidence', starting from Chapter 61.

The mathematics that follows is mostly very simple. And so is the universe - physicists and philosophers, and not only Neil Turok, often point out that the physics of the universe is beautifully, deeply simple.

Someone once said that physicists tend to work at the highest mathematical level they can. Whatever the upper level of sophistication they're capable of, that's near to where they'll be found. There's more than a little truth to this. Recently some have been searching in unnecessarily complicated places, like the idea that the universe's number of dimensions isn't a whole number - to me the clues suggest something far simpler. But this is part of a very general problem, I'd say it's a major one, and it's everywhere.

This point about levels of mathematical sophistication means that if we find ourselves in a comparatively simple universe, as it seems we do, that should affect how much time we spend on the conceptual side of physics, and how much time we spend developing advanced mathematics. And if our universe is ultimately simpler than our present abilities in mathematics, we might be looking in the wrong places.

Someone might ask why the theory here doesn't involve more mathematics. The answer is: it had often been done by others already. What was needed at times, rather than new mathematics, was to look for a conceptual picture to go behind mathematics that already existed.

Not everyone sees the importance of doing that, but during the 20th century large areas of physics remained uninterpreted. It means a new interpretation could be a new *kind* of breakthrough - a kind that we're not even used to yet, and of which we haven't yet learned the value.

But interpretation alone was never going to be enough. Efforts to find ways to distinguish between PSG and other physics, and hopefully to show PSG to be right in that way, extended the work on it by about ten years. Any theory needs to diverge from other theories, and to get to its own mathematics and ways to test the ideas, and to find its own ground.

60. Testing a mechanism

Although the mechanism for gravity I've described is new, it's not unfamiliar. It uses a process that's well understood - refraction. So apart from anything else, it's interesting because it's going to be possible to check if it's right. Not all gravity mechanisms can easily be checked. Sometimes there are too many unknowns to pin it down, and really find out what it does.

Le Sage's 'pushing gravity' is an example. It originated centuries ago, and still has a few supporters. The idea was that space has very small particles flying in all directions, and they push an object equally from all sides. But if there's another mass nearby, say a larger mass like a planet, it acts as a shield, and absorbs the particles coming from one particular direction. So there's now an inbalance, particles coming from the other direction push the object more strongly, making it move towards the planet. As it approaches, the shielding becomes increasingly effective, so it accelerates.

The idea was surprisingly difficult to disprove. Some of the parameters were unknown, such as the size and speed of the particles that did the pushing, and how they collided with things. So it was a vague hypothesis more than a theory, and when a problem was raised with the idea, it would often wriggle out of it. It dodged problems that way for hundreds of years, but during the 20th century they managed to corner it, and show that Le Sage's picture can't be what's happening.

By contrast, helical path refraction is very clearly defined. And when it comes right down to it, the way matter behaves in the large-scale world will either fit with this small-scale refraction picture, or it won't. If helical refraction and gravity do match up (and they do), that isn't a proof that helical refraction is happening. But if the mathematics describes gravity widely and well, coming from a simple starting point without further adjustments, then because the setup can also explain a range of other things, it could be seen as very strong evidence. There's also 'smoking gun evidence' in one area.

It should be understood that from here on, this is not just another feature of the similarity of the two large-scale pictures of a gravity field: curved space and refractive medium. These pictures have equivalent behaviour, although to many it's a loose similarity. RM gravity has often looked like an interesting alternative view - *but of GR, not of gravity*. That's how Eddington saw it. Still, the picture would get vague in places, and seems to fail, so standard GR with curved space seemed to win easily.

Unlike that large-scale similarity, this small-scale picture is not an alternative interpretation of GR. Instead it's a different theory of gravity. It has different equations, and they ultimately disagree with GR, around nine decimal places down the line. It fixes problems that theories of that kind used to have, and it differs from GR in a testable way. So at this point RM gravity is doing more than just interpreting GR, as it has often done in the past.

61. Removing the fog: a mathematical shortcut

The bit of mathematics that follows originally came from work on hyperbolic orbits. But it later turned out that it can be applied everywhere: to any orbit or free fall trajectory, through any spherical gravity field. If an object moves in a gravity field, here's something about its path that we didn't know about. It comes from applying to matter a law normally used for light. Its a 'smoking gun', as it shows that every point on a falling object's trajectory is connected to every other point on it, via the law of refraction.

Hyperbolic orbits are open orbits that come in from infinity, sweep past the mass, then go flying off out again. The speeds are higher than the equivalent closed orbits. At some distance away from the mass, a hyperbolic orbit is like vertical free fall. But the trajectory isn't quite vertical - when it gets there, the object misses the mass and swings around it. But as with vertical free fall, the trajectory can come in from an enormous distance. And with this kind of trajectory, a very simple rule applies.

As the object is pulled towards the mass, suppose helical refraction really is what causes it to move in that way. If so, and if only that causes it to move, it should be possible to trace what's happening using Snell's law. And it is: the refraction process can be traced all the way from infinity to periapse (closest approach). Before showing how this can be done, I need to say something about refraction, and point out a shortcut that can be used.

Snell's law can be applied at the boundary between two materials that have a different refractive index, ie. with different travel speeds for light. It uses either the refractive index of each material, or the travel speed (as a fraction of c). I'll use speeds, which is even simpler than the other version.

But if instead of two layers, there are three layers of increasing or decreasing density, or many layers, there's a simple shortcut. It's surprisingly general in what it does. I found it in 2005, and because it gave exact numbers, assumed it must be known about elsewhere. After testing it many times, I knew it had to exist already, simply as a generalisation of Snell's law.

It did - having used it for years, one day I found it in a book on the physics of sound waves under the ocean (Snell's law can be applied to many kinds of waves). Then looking more carefully through some better sources revealed it in the physics of light rays as well. So what to me had been a useful shortcut, was also just the full expression of Snell's law, generalised to cover the path of a single ray through a graded medium, or a medium with many layers. It's not a shortcut, but it can be used as one. It can be expressed as

$$\frac{sin\,\theta_1}{S1} = \frac{sin\,\theta_2}{S2} = \frac{sin\,\theta_3}{S3}$$

(5)

and so on, where S_1, S_2, and S_3 are different refractive medium travel speeds, and θs 1, 2 and 3 are the angles of a single light ray as it passes through the different layers. So here's the shortcut: with three layers, mathematically it's exactly as if the middle layer isn't there.

If you picture a light beam passing through three layers of increasing density, each layer will slow it down a little more. Say a plane flies over a lake. A light beam travels diagonally downwards from the plane's headlight: though air, into thick fog, then into the lake. Each layer is more dense than the last. The beam's angle changes twice, pointing more steeply downwards each time it crosses a boundary.

But Snell's law can be applied as if the fog wasn't there. The exit angle is the same either way. If the plane flew over on two different days, once with the fog there and once without, the angle through the water is the same on both days. And yet the layer of fog does indeed slow the light down, and changes its angle. So it might seem that Snell's law should be applied twice. You can do that, and go via the middle layer, doing two calculations, A to B, then B to C. That will give the right answer, but instead you can just take the first and third layers, A to C, which are air and water, apply Snell's law once, and the angle will be the same.

And this works just as well if instead of three layers, there are a thousand. Of all those layers, just two are needed, the first and last. To get the exit angle, all that's needed is the initial angle, the initial speed, and the final speed. The light changes direction many times, but it's as if the intermediate layers are not there.

And this also works for a smoothly graded refractive medium, where there's an infinite number of layers, so the light's path through them is curved. And it even applies if the medium is an irregularly graded one - suppose the fog density is irregular (along the flightpath direction), and the light beam curves through it. The eventual angle into the water is still the same, because all the speeds and angles stay in the same relationship.

Any light beam has its own constant, which can be derived from the emission angle θ, and the local speed for light, S, at the emission point. The number is $(\sin \theta)/S$. Once the light has set off, that number belongs to it, and if its value is known at any point on the light's journey, it can be used anywhere else on it - to get the speed from the angle, or the angle from the speed.

This means Snell's law in its familiar form can be used to relate the angles of a light beam at any two points *any distance apart*, in a graded or layered refractive medium. It's as if they're adjacent. And there's a close connection between any point on the light path and any other point on it: the law is far more general than it might seem. This aspect of Snell's law is useful, to put it mildly, in helical refraction. When I found it, I started to think it might be possible to test out the picture I'd been putting together.

62. Four numbers

In the mathematics of gravity working on matter, one doesn't expect to find Snell's law. It only gets used with light. But in helical path refraction light and matter are similar. I'm going to prove, beyond any doubt, that every point on a falling or orbiting object's trajectory behaves exactly as if it's connected, via Snell's law, to every other point on the trajectory. The derivation below is simple, but its purpose is to prove something, so I'll do it clearly - in words of one syllable, as it will probably seem to some. The aim is to show each step on the way. Snell's law relates two speeds and two angles:

$$(\sin \theta_1) / (\sin \theta_2) = S_1 / S_2 \qquad (1)$$

θ_1 and θ_2 are angles with the normal line, for a light ray crossing a boundary between two refractive materials. S_1 and S_2 are the light travel speeds in the two materials. I'm going to replace those four terms with four expressions - groups of terms - to show what causes gravity.

Because of the shortcut, instead of the border between two materials, this can be about any two points a large distance apart. And quite a few numbers turn out to make no difference: the distance between the two points, the slope of the medium's gradation, how the slope varies, and so on.

But there's just one thing: all of this only works for one single light beam. You can't relate different light beams on different paths. And the equivalent of a single light beam, as we're taking matter to be like light, is a single trajectory of an object through the field. Each unit of matter within the object follows a helical path which, when unrolled, is like the path of a single light beam. So any two points in the field can be chosen, but they must be on the same single, real trajectory. And that, along with treating matter like light at all, is 99% of why what I'm going to show you wasn't found before.

63. Towards an equation

A moving object, in this theory, has two speeds. It's best to call them a speed and a velocity, which is nearer the correct use of the terms. Any object has a light-like speed at a small scale, and a matter-like velocity at a large scale. It's true that over a distance the large-scale path curves, and strictly only a speed curves. But the small-scale path curves a whole lot more, so call that a speed. Labelling them, the small-scale speeds can be S_1 and S_2, and the large-scale velocities can be v_1 and v_2.

So you have a speed along a small-scale helical path, and a velocity along the large-scale orbit trajectory. The small-scale speed is fast, and matter 'spirals' along, covering a very much longer path. The large-scale velocity is slow, the object travels a path that (over a short distance) is a straight line. The object covers both paths in the same period of time, and arrives in one piece.

We'll choose two random points on any real trajectory. Refraction describes the light-like behaviour, so for Snell's law you put in the small-scale helical path speeds, S_1 and S_2. To get a helical path speed, you get the transmission speed at that point in the field, and see what speed the refractive medium allows light, or anything similar to light, to travel there.

The two random points you choose can be named r_1 and r_2, being at those radii. The refractive medium speeds at those points, S_1 and S_2, are given by the field's slowing factor, which is expression 4, so they'll be

$$\sqrt{1 - (2GM/r_1c^2)} \text{ and } \sqrt{1 - (2GM/r_2c^2)} \text{ .}$$

These give numbers that look something like 0.99999995. They're fractions of the speed at which light would otherwise travel, so they're speeds for light beams that have been slowed down very slightly.

But Snell's law uses the *ratio* between S_1 and S_2, and that means even more things make no difference. That includes the units. Because of that, the two expressions above can just go into the equation as they are, giving the ratio between them, replacing S_1 and S_2. So we've replaced the right hand side of the Snell's law equation:

$$(sin\ \theta_1)\ /\ (sin\ \theta_2) = \frac{\sqrt{1 - (2GM/r_1c^2)}}{\sqrt{1 - (2GM/r_2c^2)}}$$

$$(6)$$

That's the speeds done, now the angles. To check if refraction is happening, or could be happening, the two angles on the left hand side must also be replaced, and with something that can be turned into numbers. The aim is to put numbers in everywhere, then we'll see if the two sides match up, as they would do if Snell's law was at work.

So far, everything in equation 6 can be put into numbers, except the angles θ_1 and θ_2. To get those angles you need an equation from PSG that was set out earlier.

As the object moves towards the mass, accelerating as it goes, perhaps what you're really seeing is light, or something like it, obeying Snell's law. If so, the helical paths get more stretched out as the object accelerates, so the angle is changing. But in any particular place, if you unroll the path, you'd simply find a straight line set at an angle. That's why one picture translates neatly to the other. Chapter 38, p63, has the PSG formula for the helical path angle:

$$\theta = \arcsin\ (v/c)$$

$$(3)$$

That equation, which came out of the ratio between two of the three speeds on the helical path, can be used to define the lightpath in a particular place. And if refraction really is what's pulling the object, one angle you find there will be a refraction angle.

So having that equation we can substitute some terms. Instead of θ_1 and θ_2, it can be arcsine (v_1/c), and arcsine (v_2/c). But to put these angles into a form for use with Snell's law, one small adjustment has to be made: they must be subtracted from 90 degrees, to get the angle with the normal line.

I'll explain why briefly, as an aside, which only some will need. The formula that's used for translating between angles and velocities, arcsine (v/c), gives the angle between the unrolled path and the cylinder's base. That's a helpful way to express it, because a velocity of zero creates an angle of zero, and both increase when the object moves. But that's still only a convention. It's just as valid to define the helical path via another angle: the angle between the unrolled path and the direction of the object's motion. Either works, but the first is more intuitive, so I've been using that.

But the angle with the normal line, as is needed for Snell's law, is the other angle. To get from the usual angle to it, it needs subtracting from 90 degrees. The adjustment is utterly straightforward, because 90 - [arcsin (v/c)] is arccos (v/c). That angle defines the helical path just as well as arcsine (v/c).

So the two terms we need, angles with the normal line, are arccos (v_1/c), and arccos (v_2/c). So that's it, now we've got all four terms. With two speeds and two angles, it's possible to write down the equation. And then you can watch the basic law at work that many discovered one by one: Sahl, Harriot, Snell, Descartes, and probably many others. The law they found is for straight line light paths in two dimensions, not curved line matter paths in three. But it's essentially the same - one path unrolled is the other.

Putting those four expressions into Snell's law, $(\sin \theta_1)/(\sin \theta_2) = S_1/S_2$, which includes taking the sines of the angles, this leads to the statement that if PSG is right, on any free fall trajectory through any spherical gravity field, at any two points r_1 and r_2 with velocities v_1 and v_2, it should be true that

$$\frac{\sin (\arccos [v_1/c])}{\sin (\arccos [v_2/c])} = \frac{\sqrt{1 - (2GM/r_1 c^2)}}{\sqrt{1 - (2GM/r_2 c^2)}} \qquad (7)$$

This equation is still in a raw form, but it's one of the main PSG equations, as it leads to other mathematics. It can be set out in a better way without the trigonometric terms, using the fact that

$$\sin(\arccos x) = \sqrt{1 - x^2} \qquad (8)$$

Once that has been put in, you remove four square root boxes to simplify it:

$$\frac{1 - (v_1/c)^2}{1 - (v_2/c)^2} = \frac{1 - (2GM/r_1 c^2)}{1 - (2GM/r_2 c^2)} \qquad (9)$$

This is called the helical refraction equation. It turns out that this relationship *always* applies. It can be tested in about 10 minutes, by picking two numbers at random for r_1 and r_2, which are any two points on any orbit or trajectory for any object, moving through any spherical gravity field. And if the object is being 'refracted to Earth', or around it, this test should work.

And it does: the left hand side of equation 9 always gives the same number as the right hand side, to around 17 decimal places. So whatever two points you happen to choose, you find a hidden connection between them.

That's using numbers for velocities v_1 and v_2 from standard theory - from the vis viva equation, which we know is very accurate. (But naturally enough, the two sides would agree exactly if one adapts this to make a PSG orbit velocity equation - I'll do that later on.)

But if one uses numbers from standard theory, not PSG, there's a difference. It leads to a difference between the orbit speeds after 9 decimal places. And although it's a very small amount, that difference turns out to be a real one. And where there's a difference, that's not bad, it's good.

64. Testing the pattern

This equation was reached from the idea that a law found a thousand years ago is what causes the pull of gravity. It's as if Ibn Sahl's manuscript, with his famous geometrical drawing, is rolled up under his arm as he walks over to show it to someone. If so, he would accidentally have represented gravity. I'm sure he wouldn't ever do that, carelessly rolling it up, and it survived a thousand years because he wasn't a slapdash person in that kind of way. But apart from that one thing about the geometry, it's as he drew it.

Ibn Sahl's geometrical drawing described a sideways *pull* on light that makes it change direction. Although in PSG the path is rolled up, you also get a pull, and it matches how objects are pulled to Earth. The shortcut I've described means that Snell's law - it should be called Sahl's law - works at a distance. So when an object falls, any two points on its path show the pattern.

This can be checked with a calculator over a few minutes, with easily enough accuracy to prove that it works. You just translate all the terms of equation 9 into numbers. So choose two points at random for r_1 and r_2, but they must be on a real trajectory. The easiest way is to take two points on the orbit of any planet in the Sun's field (the numbers are even easier to get if you take aphelion and perihelion), where the test is accurate to 15 decimal places. In the Earth's field it's 17.

Then you use the standard orbit speed equation, the vis viva equation, to get v_1 and v_2. Using the convention with the semi-major axis a negative number for hyperbolic orbits, but a positive one for elliptical orbits, it's

$$v = \sqrt{GM\,(2/r - 1/a)} \tag{10}$$

Use that twice, to get velocities at two randomly chosen radii. Wikipedia may give the semi-major axis 'a' of the orbit, which will be the same number both times, so you get v_1 and v_2. You then have all the numbers needed to put into equation 9: two r values, two v values, and M is the mass, whether the Earth, the Sun, or any other mass. There are also the fixed numbers G and c. They all have to be in standard SI units, as with most equations.

The two sides should agree. I don't want to overemphasise it, but if you work on something for twenty years, of course you hope you'll be able to prove it, or show that it's true. But if the proof of it looks good, in one kind of way that's just par for the course. As Mel Brooks might have said, if you work on something for twenty years, it better be good.

But there's a more serious point - if you find the right picture, and play with it for years, you might find a proof of this kind. The right picture will contain things like that.

The fact that this refraction-based relationship exists supports not only PSG, but also the background theory, which gave the formula for the helical path angle. And that came out of a very different setup, put together in attempts to resolve different questions.

Those four numbers being in that relationship isn't known about elsewhere. It wouldn't be found unless one was looking for it, because it's not a general equation in one way: the numbers have to come from a particular trajectory. But that aspect makes it far more likely to be right. This connection between each point on a trajectory, and all the other points on it, is a highly specific thing, and it just doesn't look like a coincidence. It looks, when one finds that Snell's law applies so broadly there, like helical refraction at work.

65. The speed of a falling object

It's easy to calculate the speed of a falling object. (I've gone back to using the word 'speed' for large-scale motion.) It might be an everyday kind of thing, like, say, a drop of water falling off a roof. The object's speed at a given point on its trajectory, if one knows its speed at an earlier point, can be arrived at using Newtonian gravity. This is done via the acceleration, or via the energy, and people have been doing it for hundreds of years.

But these are indirect ways of getting to it, and the speed itself seems to be an indirect quantity, not a fundamental aspect of the field. The speed is very dependent on other numbers, such as how long the object has been falling, at what height it was released, and so on.

By contrast, the acceleration looks fundamental. At a given point in the field, all falling objects are accelerating in the same way, and the acceleration only depends on two things - the mass of the planet, and the distance from it. So reaching the acceleration directly is routine, but reaching the speed *directly* from the speed elsewhere on the trajectory is unexpected (though it can be done in GR), particularly if it comes out of a mechanism that tries to explain gravity itself.

Among the equations of PSG, the next is called the free fall equation. It gives the speeds of falling objects, whether they're falling vertically or sideways. If an object at radius r_1 in a spherical gravity field falls with a speed v_1, its speed v_2 at any other point r_2 on its trajectory, will be:

$$v_2 = c \sin(\arccos \frac{\sqrt{1-(v_1/c)^2}\sqrt{1-(2GM/r_2c^2)}}{\sqrt{1-(2GM/r_1c^2)}})$$

(11)

Using this, one can imagine a falling object at any point in a field, choose its speed at that point, and GM, the strength of the field. One can then calculate its speed at any other point on its fall, relating the two. And one can get *all* its other speeds along the trajectory in this way, both above and below that first point. It's clear enough how this was derived from the helical refraction equation - the two points chosen at random now become the two points to be related, and the equation is then moved around to extract v_2.

The result might seem like not much more than a restatement of the helical refraction equation, and it isn't. But it reveals a few things more clearly, and allows further exploration of the theory, as I'll show in a minute. Given what it does, it's significant that it contains c, the speed of light. It gives an object's everyday speed, and could describe any low speed event. Equations like that, for the motion of matter through a gravity field, usually don't contain c. (The adjustments onto Newton's theory do, but PSG puts in similar adjustments, and onto a basic equation that *already* contains c.)

The fact that the free fall equation contains c, and gives the right speeds, fits with the idea that matter's true speed at a small scale is the speed of light. It suggests this not just because it contains that number, but because of how it was derived, which came out of certain assumptions about matter.

The free fall equation can be set out better without the trigonometric terms:

$$v_2 = \sqrt{c^2 - \frac{(c^2 - v_1^2)(1 - [2GM/r_2c^2])}{1 - (2GM/r_1c^2)}}$$

(12)

A simple way to check it is to compare the numbers it gives with those from this standard theory equation:

$$v_2 = \sqrt{2GM(1/r_2 - 1/r_1) + v_1^2}$$

(13)

This is the direct equivalent coming from Newtonian type physics. These two equations came out of two very different starting points: energy, and helical refraction. Neither approach referred to the other on the way. But if you put in the same values for G, M, r_1, r_2, and v_1, you get values for v_2 back out from both that are surprisingly similar - they agree to 9 decimal places.

Both equations do the same trick. But they're mathematically different, and comparing them, it starts to become clear that helical path refraction mimics standard physics across a huge range of situations.

If an object is dropped one meter from the ground, the free fall equation can give you its speed when it hits the ground. It's very slightly different from the standard version. And this works for objects travelling both up and down, so if someone throws an object upwards into the air, if you know the throwing speed, you can find out the 'pause in the air' radius, or at what height it stops before falling back to Earth.

One day walking in the mountains in Arizona (I was alone there for a while, my partner had gone to New York to look after her uncle), I started thinking about what this would say about escape velocity. Experimenting with that, I found a way to prove that the height the object will stop at is always finite, if the throwing speed is less than escape velocity. But when it's exactly escape velocity, the distance becomes infinite. The object doesn't pause anywhere, it just keeps going. This led to what's arguably a more complete explanation for Newton's escape velocity equation than the existing one.

66. Escape velocity

The escape velocity equation is interpreted in standard physics already. But there's less of a complete explanation, because there's less of a mechanism for gravity. The original explanation comes via energy considerations. It was constructed from the work of Newton and Leibniz together, although they were rivals. That energy-related picture of a gravity field works, but I'd say it traces what happens, rather than explaining it.

More recently, general relativity has given a more detailed picture, but it still has a general shortage of explanation. The basic formula for escape velocity from Newtonian gravity is:

$$v = \sqrt{2GM/r} \qquad (14)$$

At the Earth's surface, that gives a value for v of 11186 m/sec, and we have to throw things off the Earth at higher speeds than that if we want them not to fall back down again, which means you have to use a rocket.

From equation 12, the free fall equation, the first thing is to get an equation for the height at which any object thrown upwards will pause before falling back down again. If you know the throwing speed v_1 and the Earth's radius r_1, you can work out the point at which objects moving upwards will pause in mid-air, which can be r_2. It's where $v_2 = 0$.

Translating some terms from equation 12, the upward throwing speed and the place where the throwing is done, v_1 and r_1, become simply v and r_E, the Earth's surface radius. The stopping point in mid-air can be r_{MAX}.

Now v_2, the speed at r_{MAX}, being at the pause point, is zero. When the object pauses, the helical paths will become horizontal circles, therefore the angle between their path and the normal (a radial line) at that place will need to be 90 degrees.

And you find that it is. Looking at equation 9, the helical refraction equation, the angle with the normal at r_2 is arccos (v_2/c), and because $v_2 = 0$, you get arccos 0. And as arccos 0 = 90, the angle is indeed 90 degrees. The complete term from equation 9 is the sine of the angle, sine (arccos $[v_2/c]$), and as the sine of 90 is 1, the term disappears, as it should with a speed of zero. These points help to show that the PSG picture is self consistent, just as if refraction really is at work.

The 'pause in the air' equation, for the maximum radius of an object thrown upwards, is:

$$r_{MAX} = \frac{(2GM/c^2)}{1 - (\,[1 - (2GM/r_E c^2)]\, / \,[1 - (v^2/c^2)]\,)} \qquad (15)$$

For the pause height, you get r_{MAX}, then subtract r_E. It was quite fun testing this out, for instance, you can imagine a soccer penalty kick is taken vertically and find out how high it goes. The speed of a penalty is about 70 mph, and it would travel up about 45 yards.

To test Newton's escape velocity formula, make the throwing speed escape velocity, then ask the 'pause in the air' equation what happens. You put the square root of $2GM/r_E$ into equation 15, to replace v. You find there's a large cancellation of terms, and the end result is:

$$r_{MAX} = (2GM/c^2) / (1 - 1) = \infty \qquad (16)$$

So the object pauses in the air at an infinite distance away, which means that it never does, so it escapes. This shows that PSG reproduces escape velocity, another of the well known, well tested aspects of gravity. According to PSG, there's a definite limit because of the basic nature of refraction, as with the bouncing laser. The equation we use for this limit is what it is, because that cancellation of terms shows the point where the refraction no longer has a turnover radius.

Incidentally, when the throwing speed is escape velocity, you get a parabolic orbit, and Newton's theory and PSG converge even more than usual, as they give identical numbers. Equation 13, from standard physics, also does what equation 12 (the free fall equation) does. It leads to a setup that explodes to infinity when you put in Newton's escape velocity equation. But PSG explains it better, as it describes it via refraction, which is a well understood process, including aspects like the turnover point, which we know refraction does.

67. One equation

What all this leads to is a single equation. It's a speed equation for all orbits, and all the different possible trajectories through any spherical gravity field. It does something similar to the vis viva equation from standard theory, but it's different, and according to PSG, more accurate. Accuracy is often related to scale. If the Planck scale really does look like a lot of parallel cylinders with circling waves on them, obeying simple rules, an equation that comes out of that could be as accurate as it gets. So working on this, to me it was about trying to get to a very fundamental equation.

The vis viva equation came from a theory by Leibniz in the 17th century, that I've mentioned. Initially it rivalled Newton's work, but later on the two were seen as partly complementary. Much of the vis viva theory disappeared, or was absorbed into the more recent view of energy. By contrast Newton's work remained undiluted, and stood alone for centuries.

But Leibniz's theory was an important step, and it did produce one key thing that remained unaltered: our main orbital speed equation. It's very general, and works for all orbits, whether elliptical, hyperbolic, parabolic or radial. It's so accurate that NASA still uses it constantly. As Leibnitz was self-taught in mathematics (Chapter 111), it was lucky he managed to publish his work at all. Using the more common convention, with the sign of the semi-major axis varying depending on which kind of orbit it is, the vis viva equation is:

$$v = \sqrt{GM\,(2/r - 1/a)} \qquad\qquad (10)$$

This was reached via the energy of orbits. It turns out that a similar equation can be reached from PSG and Snell's law. To get to it, the helical refraction equation is adjusted through a series of steps until something resembling the vis viva equation comes out. Although superfluous for some, I'll show more of the stages on the way than one might, because the derivation is part of what shows the point I'm making.

There are two alternative conventions with hyperbolic and elliptical orbits. The standard one has the sign of the semi-major axis switching between plus and minus, depending on the type of orbit. It's positive for elliptical orbits, negative for hyperbolic ones. This allows there to be just one version of the vis viva equation, as above.

But for what follows, the other convention will help simplify things, in which the semi-major axis is always positive. A page or so from here we'll go back to the more standard convention, and use it from there on throughout. The helical refraction equation will be the starting point, and it started out being about hyperbolic orbits:

$$\frac{1 - (v_1/c)^2}{1 - (v_2/c)^2} = \frac{1 - (2GM/r_1 c^2)}{1 - (2GM/r_2 c^2)} \tag{9}$$

Now it needs to go through some changes. There needs to be just one point chosen at random, not two. So one of r_1 and r_2 has to be a fixed point, while the other can be any point on the orbit.

So r_1 and v_1 get fixed values, and go into the algebra, while r_2 and v_2 become simply r and v: any point on the trajectory and the corresponding speed. To fix r_1 and v_1, the radius r_1 is made infinite, and v_1 will then be the hyperbolic excess velocity

$$\sqrt{GM/a} \tag{17}$$

which is the speed a hyperbolic orbit approaches as it moves away to large distances. That speed is the same in PSG as in standard theory: a number of terms disappear at infinity in both. So all the other speeds on the trajectory can be related to that. The next bit is just about juggling algebra, so partly to differentiate, I'll leave the steps on the way in the old equation font, as was originally used (in the last part of the book it's also used in the same way, for the less important equations).

To start, equation 9 is set out with $r_1 = \infty$. Snell's law turns out to apply even across a hypothetical infinite distance of graded refractive medium:

$$\frac{1 - (GM/ac^2)}{1 - (v/c)^2} = \frac{1}{1 - (2GM/rc^2)} \tag{18}$$

The 1 on its own appears when r = ∞, because the refractive medium thins out completely there, so the RM slowing factor goes to 1, from just below 1. From there it's just a case of simplifying, and extracting v. I'll show some of the stages on the way, as the equation changes:

$$1 - (GM/ac^2) = \frac{1 - (v/c)^2}{1 - (2GM/rc^2)} \tag{19}$$

$$1 - (v^2/c^2) = [1 - (GM/ac^2)][1 - (2GM/rc^2)] \tag{20}$$

$$v^2/c^2 = 1 - ([1 - (GM/ac^2)][1 - (2GM/rc^2)]) \tag{21}$$

$$v^2 = c^2 \{1 - ([1 - (GM/ac^2)][1 - (2GM/rc^2)])\} \tag{22}$$

$$v = c\sqrt{1 - ([1 - (GM/ac^2)][1 - (2GM/rc^2)])} \tag{23}$$

$$v = \sqrt{(2GM/r) + (GM/a) - (2[GM]^2/arc^2)} \tag{24}$$

and the last simplification gives:

$$v = \sqrt{GM(2/r + 1/a - 2GM/arc^2)} \tag{25}$$

This was for hyperbolic orbits, but with the semi-major axis positive. Now we can switch to the more standard convention, allowing a single equation for all orbits, and use it from here on, throughout. So for hyperbolic orbits, the semi-major axis 'a' is now negative, but for elliptical orbits it's positive. That means switching the signs in the equation:

$$v = \sqrt{GM(2/r - 1/a + 2GM/arc^2)} \tag{26}$$

And that's the PSG orbital speed equation. It works for all orbits and free fall trajectories, and calculations have shown that (in the common convention, where the sign of the semi major axis varies) the third term in the brackets always has a plus sign before it.

So there's now a general PSG equation that has the same role as the vis viva equation, and resembles it, but has a very small extra term. And it turns out not to be a relativistic correction - I'll put those in later - or anything else that can be explained away. Instead, if PSG is right, it's new physics. What that extra term does is to make open orbits (which are fast) slightly slower, and closed orbits (which are slow) slightly faster. It brings the two kinds of orbits a little closer in speed than they are in standard theory.

As always, what interests me is the differences between PSG and standard theory, but the extra term is small. In the Earth's field, speeds from PSG and standard theory diverge after nine decimal places. But they do, which means the vis viva equation may only provide approximations to real orbits - helical refraction orbits may be more accurate.

This difference is small enough to have avoided being noticed, in the inner solar system. Some of it disappears into differences to parameters that have their values estimated, because they can't easily be got at in any other way. We generally estimate the masses of planets like Earth or Mars by assuming standard theory is right, and then calculate the masses from how things orbit around them. But if the orbit speed equation that's used is wrong, the mass value arrived at may also be slightly wrong.

The parameters that are estimated via GR are part of a closely knit network of interrelated numbers. Sometimes there's no way to know what the data is telling us, and whether there are alternative interpretations for it. The solar system might look much the same either way.

68. Some implications

RM gravity has always been better at the behaviour of light than matter, and better at getting things to go around a mass than straight towards one. It's harder to make an object fall vertically (though they do), because there don't seem to be any angles involved, and refraction is about angles. But in PSG there are hidden angles: the paths that matter takes, although very thin and tightly wound, have angles that can be calculated. And an angle corresponds to a speed, which makes it possible to get a handle on things.

The result, the helical refraction equation, shows that the picture PSG came from, as well as interpreting quantum mechanics, also generates gravity. As I pointed out, orbital speeds contain various other elements of gravity, which are implied in them. And it's not just *some* orbital speeds, it's all of them. Being able to generate that entire set of numbers also means being able to generate gravity.

Although the PSG orbital speed equation gives numbers close to those from the vis viva equation, and resembles it, the two equations are unrelated. The vis viva equation was not needed to reach it, and on the face of it, one could reach that equation via helical refraction, and not much else.

But that's at first glance. In fact there's the gravitational redshift, which led to the transmission speed. That's the only direct mathematical link to other gravity physics, but it's a major one. But other theories provided a template, pointing the way in all kinds of ways. Then there are indirect links to the RM gravity work of Newton and Eddington, and other non-gravity physics that affected the background theory. So connections with existing physics are in many places, and in reality - it hardly needs saying - without other theories, PSG would never be found in ten thousand years.

69. Deriving the force of gravity

As I said in the section on the force of gravity, the equivalence between mass and energy, discovered centuries after Newton's force equation, means the gravity equations can be set out in two different ways - describing gravity via an object's mass, or its intrinsic energy. The main PSG equations for the force of gravity, and for the acceleration, go via an object's mass in a conventional way, and are at the end of this chapter. They're the important equations (32 and 33), but this energy approach is still worth setting out.

I've put forward the idea that what actually causes the force of gravity is the local rate of change of the transmission speed of space. This quantity goes by an inverse square law, so it fits the bill in that way. It also fits the bill in the conceptual picture from PSG of how gravity works. And here I'll show it also fitting the bill in another way: in one way to reach the force of gravity in PSG, this local rate of change is simply multiplied by the object's intrinsic energy. What comes out, whether or not an approximation, is Newton's well-tested equation for the force of gravity.

The rate of change can be expressed as dx/dr, or 'the difference in x, over the difference in radius'. It shows the way the transmission speed is changing in the radial direction, at a given place in the field. The x represents expression 4, as was set out before:

$$\sqrt{1 - (2GM/rc^2)} \qquad (4)$$

So x is the transmission speed of the background space, expressed in terms of c. It's a speed, so call it v, the complete term can be $dv/dr(r)$. The (r) shows this is at a particular radius. To get the force at work, F, you take the apple's intrinsic energy E, or mc^2, and simply multiply it by that rate of change:

$$F = E \frac{dv}{dr}(r) \qquad (27)$$

It turns out that this is the same as Newton's:

$$F = GMm/r^2 \qquad (28)$$

To show that equations 27 and 28 are the same, *dv/dr* can be approximately expressed as:

$$GM/r^2c^2 \qquad\qquad (29)$$

which includes the central mass, so it quantifies the strength of the field. In PSG it gives the rate at which space is changing at that location (in standard physics it's the rate of change of some other things about the field).

So we have $F = E$ (*dv/dr*). Both terms on the right hand side, E and *dv/dr*, can be expressed differently: $E = mc^2$, and $(dv/dr) = GM/r^2c^2$.

So the trail to Sir Isaac's work is: $F = E$ (*dv/dr*) $= mc^2$ $(GM/r^2c^2) = GMm/r^2$. The inverse square pattern that according to PSG leads to the inverse square law can be seen, sitting there in the algebra - that's just what *dv/dr* happens to come to, when converted to that kind of expression.

But so far this has been oversimplified. It can be made a bit more precise by defining some terms more accurately. The rate of change of the transmission speed of space, *dv/dr*, rather than being exactly GM/r^2c^2, is in fact:

$$(GM/r^2c^2) \; / \sqrt{1 - (2GM/rc^2)} \qquad\qquad (30)$$

which is expression 29 divided by a number very slightly below 1 (expression 4 again), leaving it almost the same. This very small difference means, on the face of it, that the PSG equation seems only to approximate the Newtonian equation. But in fact it adjusts things in an interesting way.

This approach from PSG uses the apple's intrinsic energy, and we know from the gravitational redshift that an object's energy effectively varies slightly at different heights in the field. There are different views on this, and no clear cut explanation. But it's there in the mathematics: frequency is proportional to energy, so as frequency varies with radius, so does an object's energy. Not everyone thinks this set of changes is real, but in the background theory that PSG came from it's real. (Book III shows how the slowing of the transmission speed neatly explains the *reduction in energy* we find in a gravity field, which is a mystery in other contexts.) And it turns out the extra term in expression 30 adjusts the PSG force equation so it fits Newton's equation more closely, for any particular location in the field.

This way of reaching the force, using the energy, stays the same:

$$F = E\frac{dv}{dr}(r) \qquad\qquad (27)$$

But the terms are very slightly different: E in this context is mc^2 multiplied by expression 4, and dv/dr turns out to be GM/r^2c^2 divided by expression 4. So the two expression 4s cancel, and again one gets Newton's force equation:

$$F = mc^2 \sqrt{1-(2GM/rc^2)} \ x \ (GM/r^2c^2) / \sqrt{1-(2GM/rc^2)} \qquad (31)$$

When the central mass is spherical, dv/dr equals expression 30. But that will not necessarily be so if the mass is irregularly shaped, as with some asteroids we've been landing robots on recently. People have had good reason to work on modelling irregular gravity fields (there's a paper on it in the reference section). In the context of PSG, the equivalent might involve estimating or deriving some numbers for dv/dr for any point in the field.

That's because whatever the mass's shape, and the corresponding shape of its emission, it makes no difference. That way of reaching the force from PSG still applies, via the local rate of change dv/dr, and v is still the transmission speed of space. Whatever the pattern across a wider area, the resulting local force of gravity, via dv/dr, remains the same.

I'm not saying that with gravity the mass term is irrelevant, and that only the energy counts - far from it. Newton's equations show $F = ma$ at work, which has been part of the thread of the argument in this book. Either the mass or the energy can express the force, but in PSG so far the energy version is less explored. I'll say more in Book III, but it's interesting, as it shows the rate of change neatly fitting into the mathematics of gravity.

But meanwhile, rather than comparing the two approaches, I'll give the more conventional version, expressed via an object's mass, which is more rigorous, more important and relevant, and is not an approximation. The official PSG equation for the force of gravity, derived from the orbital speed, is:

$$F = GMm/r^2 + 2(GM/c)^2 m/r^3 \ . \qquad (32)$$

And the acceleration due to gravity is :

$$a = GM/r^2 + 2(GM/c)^2/r^3 \ . \qquad (33)$$

These are expressed as gravity equations are, without the post-Newtonian adjustments. But incidentally, as an aside: with the adjustments in, instead of the small added term, you have an even smaller subtracted term:

$$F = GMm/r^2 - 4(GM)^3 m/(rc)^4 \ , \qquad a = GM/r^2 - 4(GM)^3/(rc)^4 \ .$$

70. The 'light and matter at right angles' experiment

I'll return to the point about light and matter discussed in Chapters 33 to 35. Although it's only about a thought experiment and a calculation, it strongly supports PSG. Light and matter are released in a gravity field at the same time, and from the same height, with the light emitted horizontally and the matter released to fall vertically. According to PSG, ignoring post-Newtonian adjustments and the curve of the Earth in the calculation, if they both reach the ground due only to gravity, they will do so at exactly the same instant.

This is because in PSG the two paths are essentially the same, and identical mathematically in various ways. Matter follows a 'rolled up' path, while the light follows an 'unrolled' one. The unrolled path produces the behaviour and mathematics of refraction, while the rolled path produces the behaviour and mathematics of gravity. But each produces both.

Standard physics will give a very similar result, although unlike PSG, it might not have a good explanation. But this is clearly a general aspect of gravity, and it needs an explanation, because it applies very broadly - regardless of the size of the central mass, and the height at which the light and matter are released (if they both reach the ground).

With the PSG equations that follow, the Earth's mass and radius are referred to, but it can in principle be any mass. PSG, including Snell's law, holds that when the light hits the ground, the angle φ to the horizontal is:

$$\varphi = \arccos\left[\sqrt{1 - (2GM/r_2 c^2)} \,/\, \sqrt{1 - (2GM/r_1 c^2)}\right] \tag{34}$$

r_1 is the distance from the centre of the Earth to the light's emission point
r_2 is the radius of the Earth at that location
G is the gravity constant
M is the mass of the Earth

The time of flight t for both light and matter to reach the ground is:

$$t = r^2 c \, (\sin \phi) \,/\, GM \tag{35}$$

where r is the Earth's radius. The height h from which light and matter are released, $r_1 - r_2$, will be a distance small enough for the acceleration due to gravity to be taken as the same as at the Earth's surface:

$$h \approx r^2 c^2 \, (\sin \phi)^2 \,/\, 2GM \tag{36}$$

The distance d that the light travels before it and matter reach the ground is:

$$d = r^2 c^2 (\sin \phi) / GM \tag{37}$$

or alternatively

$$d \approx 2h/\sin \phi . \tag{38}$$

It's worth pointing out that doing this calculation without post-Newtonian adjustments, the result is still relevant. They make a difference so small that it's negligible. And Newton's theory is part of general relativity, and a large part of our description of gravity in the 21st century. It's more than 99.9% of it, looking at the numbers. The reason many gravity theories make miniscule adjustments to Newton's theory is that it describes gravity well already, on its own - and contains some basic clues about how gravity works. And that includes hidden clues like this one, which although a few may know about it, have remained generally unknown. Standard physics gives people no reason to be looking for it, particularly with the right angle involved.

But standard physics nevertheless confirms it. In standard physics, treating a light beam as a projectile produces more or less exactly the same numbers as PSG, and brings this question near the area I've called the uniform response of matter to gravity, which was first discovered by Galileo.

One part of the uniform response is seen in projectile motion. Galileo found that the downwards acceleration due to gravity is separate from the motion resulting from throwing the object. In the standard mathematics of projectile motion, the time of flight t, whatever the initial angle to the horizontal θ, is:

$$t = (v \sin \theta + \sqrt{[v \sin \theta]^2 + 2gh}) / g \tag{39}$$

where:
v is the initial velocity
h is the initial height
g is the acceleration due to gravity at or near the Earth's surface.

The starting velocity v is often needed, but in the special case of throwing in a horizontal direction, which means the angle $\theta = 0$, some terms disappear, and they take the v term with them. We're then left with:

$$t = \sqrt{2gh} / g \tag{40}$$

This means the starting velocity makes no difference, and the time of flight is always the same, with projectile motion that starts horizontally from a given height. So that may include events in which the starting velocity is c, such as the emission of light. One particularly good reason to think this is true is that the numbers from PSG for the horizontal emission of light, from the time of flight equation (eq. 35), agree to eight decimal places with the numbers that come out of this equation (eq. 40).

These two results that agree closely came out of very different areas: Snell's law and standard projectile motion. The term ϕ in equation 35 was derived by applying Snell's law to a light ray that at first travels horizontally. It's well known that refraction can be used to derive the effect of gravity on light. But it's far from well known that it can be used to derive projectile motion for matter.

From this it's clear that in basic gravity, light can be treated as a projectile, and light and matter sent off at right angles hit the ground at the same time. This arises in both standard theory and PSG, and is very general - enough to need an explanation. It can't be explained via curved spacetime, because the uniform response, which it and projectile motion in general imply, happens in exactly the same way in other areas (via $F = ma$), including areas where no-one is trying to suggest that curved spacetime is the cause.

And the other question, of why light should turn out to be an extra part of this uniform behaviour, is also arguably much better answered in PSG than in standard physics. PSG explains it via a close link between light and matter, and the paths they travel to reach the ground - which are exactly the same at any given moment in speed, height, and the angle to the horizontal.

Apart from PSG providing a better explanation, this involves an experiment that might actually be done. An object should fall to the ground in the same time as light bouncing between two mirrors, allowing for imperfection in the details of the bounces if necessary. One good thing about this way of doing it is that you don't have to start so near the ground, because the light is going to hit the ground anyway. That would probably be the simplest way to get at the small differences between the two theories which, when this is looked at closely, are likely to be there.

Part 13. Proof that an alternative picture exists

71. The geodetic effect

Before we get to the places where PSG and GR diverge in an observable way, I must tell an odd story, involving a situation that looked like it might do one thing, but which eventually did another instead. It then unexpectedly led to good mathematical evidence for PSG.

Like any theory, PSG has to go through a number of tests. One of them that it has already passed is explaining why the geodetic effect has been measured. As of around 2008, any theory of gravity has to explain why this effect exists. That can be a problem for a group of theories known as flat space theories, in which space is flat at a large scale, and not curved. They include PSG, and as far as I know, no other flat space theory has been able to explain this with a mathematical basis, but PSG did it recently.

The geodetic effect is a slight tilting of orbiting objects as they travel around a mass. It was first predicted from GR, and ninety years later measured from a NASA space probe. To tell if something is tilting slightly, one way is to use a gyroscope. They're used in navigation because if they're well shielded, the spin axis always stays at the same angle - unless something very fundamental is affecting them. Gravity Probe B was launched in 2004 with four gyroscopes on board. As the probe orbited, they slowly tilted further and further from their angle at launch, and by more or less exactly the right amount.

The experiment had problems, but it certainly measured the geodetic effect, and we now know it exists. What makes this interesting is that it had widely been thought that there was no other explanation for the effect apart from curved space. Some see it as curved space's 'signature effect'. So to some, when the geodetic effect was measured, it was almost like direct observation of the curvature of the space around the Earth. If PSG is to stand up as an alternative to curvature, it must give an explanation for this. And because some saw the measurement as 'proof' of curved space, it was essential that the explanation had a mathematical basis.

General relativity, after a long struggle, was completed in 1915, presented to the Prussian Academy of Sciences that year, and then published in 1916. This has caused minor confusion about the date ever since, on top of the wider confusion caused by the beautiful, complicated theory itself. Once it was out people responded quickly to it, and in 1916 and 1918 two extra predictions were published by other physicists. They involved effects that would only be found after a century of progress developing measuring instruments.

The geodetic effect arises from curved space because of a simple principle. If you slide a ruler sideways over a curved surface, keeping its position steady, or 'parallel transporting' it - it will still change its angle. This angle change can potentially reveal the curve of the surface it moves over, even if there's no other way to tell.

One can imagine sliding the ruler over a globe, keeping it parallel with where it was a moment ago. It will change its angle in two dimensions because the globe is curved into a third direction. The same principle can seem to show whether the space near the Earth is curved. Space is three-dimensional, but it might be curved into a fourth.

A gyroscope, like the ruler, always points in the same direction, or tries to. Once its spin axis is pointing in some direction, it keeps on pointing there. So if the gyroscope is moved sideways, the straight line of its spin axis is being parallel transported. If placed in a space probe and carefully shielded, most effects can't get at it, which means that only something deep, like a curving of the space it travels through, will alter its angle.

The geodetic effect was found soon after Einstein's 1916 paper, by the well-known Dutch physicist Willem de Sitter (who later worked with Einstein). He found the effect by studying the orbits of the Earth and Moon, and adding in relativistic effects. No-one thought about gyroscopes at all for a few decades, as there was no possibility of an experiment at the time. But later people realised that de Sitter had been using the Earth-Moon system as a kind of big gyroscope.

Gravity Probe B took longer to prepare than any experiment ever has. It took forty years, but they kept on getting funding. The idea was first put forward in 1959, and physicists and engineers then had to push the envelope in many areas, developing new technologies to make the idea possible. Photographs of the team working on it show their clothes and hairstyles changing as time went by. When the probe was finally launched in 2004, it was full of highly advanced technology. Each of the four gyroscopes is a quartz sphere, more perfectly rounded than any other existing human made object, except the other three. They were extremely well shielded, kept at a low temperature, then spun up and monitored for the slightest change.

It took until 2011 to get the final results. The experiment was controversial due to unexpected problems, and failed to reach the accuracy that had been hoped for. The problems increased the time it took to analyse the data to five years, and the team fought for funding to keep it going. The probe was looking for two effects - the result for the smaller frame dragging effect was also there in the figures, but more questionable.

Though extracting the final result was difficult, they eventually removed the background 'noise', and retrieved a fair amount of data. The result had the geodetic effect, and so seemed to support GR. Some have argued that the problems mean the result wouldn't have carried much weight if it had shown anything unexpected, so to them this expected result is also questionable. This is true, and it does weaken the findings overall, but to me it only affects the borderline elements of them. At the borderline where things are unclear, if a surprising result would lack credibility, then an expected one should lack credibility too. But the geodetic effect was still found, and as far as most of us are concerned, the effect has now been measured. It was not borderline at all - it was clearly visible in the data.

The other effect, frame dragging, makes a much smaller shift of angle, about 160 times smaller, and at 90 degrees to the first, due to the Earth's rotation affecting the space around it. In the final result they eventually published it was also there, and was probably measured as well. But in removing that much unexplained 'noise' they used mathematical techniques that contained significant assumptions. No-one knows what created the problems.

There are things in PSG that might have done. If so, what they removed was gravity data they had been trying to get. The geodetic effect is about a slow rotation of each gyro, while it also rotates rapidly in another direction. But according to PSG each gyro, when it tilts, is slowly rotating around a point slightly below its geometrical centre. Although this difference is very small, it wouldn't have helped.

It's possible that a reanalysis of the data would support PSG, and later I'll say more about some unknown effects that were found. But the final conclusion, after fifty years of work, was: the frame dragging result was less certain, but the geodetic effect was there. And that's the important measurement. The details of the result have bearing on another question, and one that we need to answer: the question of whether space is curved or flat.

72. Correcting predictions

One early prediction Einstein made from special relativity was wrong. It later turned out that only the prediction was wrong, not the theory from which it came. So ten years later, when the theory had been worked through in more detail, and some adjustments had been made, the right prediction emerged.

Something similar happened with PSG, except that no further adjustments to the theory had to be made in between. The correct prediction later emerged straight from the original premiss (light and matter being slowed in a gravity field by the factor that gives the time rate). It had simply been a case of the physicist being wrong, not the theory.

After the geodetic effect appeared in the data, a closer look at the theory led to a good mathematical explanation which, when published in 2008, became a central bit of evidence for PSG. It also seems to have been the first time a particular thing had been done.

Einstein's wrong prediction was that a clock placed at the North pole would run faster than a clock at the equator. Time dilation means that comparing two clocks, the one moving faster will run slower, and the one moving slower will run faster. So special relativity led Einstein to think the Earth's rotational motion at the equator, where objects move at 465 m/sec, doing a full turn around the planet in 24 hours, would make a clock there run slower than one at the pole. After all, the one at the pole just sits there turning slowly around on itself, at half the speed of its hour hand.

In fact, this published prediction was wrong, because the other kind of time dilation, discovered later, takes things in the other direction, and more than cancels the kind that's due to motion. Later, when Einstein had generalised the theory to include gravity, it turned out that gravitational time dilation, which is a separate effect, made the clock at the equator run slightly faster overall.

Due to its motion, the clock at the equator runs slower than the polar clock by the factor $1 - (1.2 \times 10^{-12})$. But the clock at the equator is further from the centre of the Earth, because the Earth's geoid shape has polar flattening and equatorial bulge. So gravitational time dilation makes it also run faster by the factor $1 + (2.33 \times 10^{-12})$. It was a close run thing between the two effects, and by chance they nearly cancel. But combining them, the earlier prediction is reversed, and overall the clock at the equator runs faster.

Einstein also corrected something else: the deflection angle of light. For light passing the sun, he thought early on that his theory gave the same deflection angle as Newton's. But ten years later, when he had generalised the theory, the predicted angle was doubled. One commentator points out that Einstein was lucky, as all expeditions to photograph a solar eclipse between 1907 and 1919 were rained off, and by the time Eddington got a bit of decent weather, relativity theory was complete.

These alterations were no problem for relativity, for a number of reasons.

But it's simpler if only the prediction needs changing, without any alterations to the theory from which it came. It's then simply a case of showing that the prediction was wrong, and that the theory in fact predicts something else. I say 'simply', but it wasn't. It took 18 months of struggle - first to solve the mathematical problem, then to get the paper into a peer reviewed journal, then to persuade people to look at it, and give the acknowledgement needed to set things straight.

What happened with PSG was about the geodetic effect. In 2007 the initial set of results from Gravity Probe B were suddenly about to be announced. I'd known about the experiment for longer, but I didn't know when the results would come though, and was very wrapped up in something else at the time. I hadn't thought about any bearing the experiment might have on my theory, as the gravity part of it was still at an early stage. I suddenly found out there was a short window of time to look at the theory and try to work out what it predicted. And looking at it, it seemed likely that with the refractive medium to replace curvature, a gyroscope parallel transported through a gravity field wouldn't change its angle.

The other effect, frame dragging, did seem likely to exist, as the RM around a rotating mass like the Earth could certainly drag things with it, just as curved space drags things with it in the equivalent picture. In PSG there are waves flowing out into space from the edge of a rotating mass, and they should pull anything nearby around with them a little.

But the geodetic effect, at that point, seemed unlikely to exist. So I hurriedly wrote a paper, which included a correct prediction about interferometers, and an incorrect one about Gravity Probe B. And thanks to the support of a well known astronomer, Halton Arp, and a recommendation he gave to the editor of a journal after reading the paper, it got through peer review just in time - two days before the results were due to be announced.

The day before they were announced, a British physicist I knew, from nearby in Surrey, said in an online discussion that he found it so exciting he was off to the conference in Jacksonville, Florida, to hear the results announced for himself as they came through. I thought he was joking, but he jumped on a plane and went over there.

The initial result from Gravity Probe B showed something that looked like the geodetic effect. There was a way to go with the data analysis, but it wasn't hard to see the geodetic effect there. Because I was absolutely sure that PSG was right, and for very well supported rational reasons, I then searched for something that could cause the geodetic effect. I never thought of changing the theory in any way, but simply searched using the original straightforward premiss - of the way in which the refractive medium slows light and matter - and the conceptual picture that went with it.

And several weeks after the results came through, an idea that I'd had while holding onto the side of a swimming pool led to a calculation. It quickly gave the right numbers every time, to many decimal places. One thing that helped with finding it was being absolutely certain that it was there to be found. The idea was this: as the gyroscope moves through the refractive medium, one end of its spin axis is nearer to the Earth, and the other end is further away. That means the two ends are moving at slightly different speeds, because the RM slows matter, including gyroscopes, in a way that varies with distance from the centre of the field, with more slowing nearer the mass. So as the probe orbits, the top end of the gyroscope's spin axis is moving faster than the bottom end. And that turns the gyro through an angle.

As the gyro is only about the size of a ping pong ball, it had never occurred to me that its upper and lower edges being at different heights could make any difference. But the angle the gyro turns through is small - it accumulates so slowly that it's usually expressed as an angle per year of orbiting the Earth. The new idea for what might cause it was in fact similar to an unfinished and inconclusive calculation from some years earlier about the Moon, to see if it turned slightly as it orbited. The Moon is so large that its upper edge might travel significantly faster than its lower edge. But the theory was still at an early stage, and at the time I had no idea that what I was looking at was an alternative approach to the geodetic effect.

The calculation for Gravity Probe B was simple. You start with a basic orbital speed across the whole object, and then add slight speed corrections for the different slowing effect of the refractive medium at different heights. It turns out that this slowly tilts the gyroscope around in just the right direction, and through just the right angle. It creates an alternative version of the effect. In both, the change of angle accumulates over time, and the PSG version would look exactly like the GR one, in the data. It would be hard to tell which was being measured.

The calculation gave the 'curvature component' of the geodetic effect - ie. the 2/3 thought to be caused by space curvature. (The other 1/3 has possible causes that exist whether space is curved or flat.) So the calculation showed a clear alternative interpretation for the curvature part of the effect, arising directly from the simple published premiss of the original PSG theory, with no changes at all. I knew that this could vindicate the theory, so I put the calculation in a paper, but none of the journals I submitted it to wanted to publish it - the paper kept getting turned down.

Then a few months later something better than a calculation came out of the idea: a generalised equation, which gave the change of angle that's caused in PSG, for a single orbit around any spherical mass, any distance from it. That got through peer review and into a journal. It was a way of generating the geodetic effect, but coming directly out of a flat space theory.

The paper, 'A derivation of the geodetic effect without space curvature', was published in The Journal of Gravitational Physics: received 21.1.'08, accepted 28.2.'08. The journal later went offline, although physical copies remained. Some authors still refer to it on arXiv preprints as their journal reference. The paper contained one equation: it was a clear alternative to the very different equation from GR, which does exactly the same thing.

73. The geodetic effect derived from flat space

As the data analysis results from Gravity Probe B kept on coming in, several websites would list the theories that were still compatible with the results, from an earlier, longer list of possible theories. The data analysis took ages, and one respectable site kept an ongoing account of it over five years, with a reliably updated list of theories that were still viable in the light of the data, and of theories that were excluded.

Many of the physicists behind these theories had met on the discussion site before any results came through. So we followed events together, chatting cheerfully about things from all over the planet, as you do, with a friendly spirit that completely ignored the fact that we would get listed there rather like horses lining up for a race.

The weeks of discussion leading up to the announcement of the first results were great fun, and it made a change from working alone. I think quite a few others felt it too - physics is sometimes a lonely thing, particularly on the theoretical side. Being with a bunch people who shared a similar quest felt great, and at times reminiscent of the fun side of school days. Then the first results came in, and PSG was quickly taken off the list of viable theories. It felt miserable being excluded, and that feeling went on.

But then the second paper was published, with the equation that proved PSG gives a similar prediction to GR after all. At that point the physicist in charge of that thread, and of the list of theories, congratulated me on getting the paper through peer review, having read it in the journal.

He checked the equation, and several other good physicists from the site's community also checked it. PSG was then reinstated, having been through a gap of 18 months, and put back on the 'viable' list of theories. It now gave the same prediction as GR for this particular set of measurements, as other theories did, to the accuracy of the experiment. It didn't matter that GR and PSG couldn't be told apart via this experiment, the main thing is to show mathematically what the theory predicts, and why. On October the 2nd '08 they put PSG back on the viable list.

If the mathematics hadn't clearly vindicated the theory, their reaction to my attempts to get PSG reinstated would have been different - to put it mildly - and before they actually saw the paper, some were aggressively telling me to 'give up', as I tried to tell them about the calculation. Some of that discussion is still online, including the early bad reactions, and later good reactions and comments from physicists who had read the paper.

And the equation itself, of course, remains as well. Mathematically the two equations, from GR and PSG, are not the same at all, including the terms. But although they're very different, the numbers they give for the tilt angle per orbit are the same for the first 8 decimal places - the difference is too small for Gravity Probe B to distinguish between them.

There are no other well-known alternative equations for the geodetic effect, so for what it's worth, it may be the first mathematical alternative to curved space, as an explanation for the effect. The reason it wasn't found before is probably the unexpectedness of the idea that matter behaves as light does in that situation, and will be slowed by a straightforward factor difference from the speed at which it would otherwise travel.

In PSG any orbiting object turns slightly in the refractive medium given off by the spherical mass it orbits, due to a local slowing of matter. For a spherical object such as one of the Gravity Probe B gyros, the angle of precession θ in degrees that accumulates over a single circular orbit is:

$$\theta = arctan \frac{2\pi r \left(\sqrt{1 - (2GM/r'c^2)} - \sqrt{1 - (2GM/rc^2)} \right)}{r' - r} \tag{41}$$

where r is the distance from the central mass to the orbiting object, and r' is the distance to a point on its upper edge (equal to r plus the object's radius). G, M and c are as you'd expect. This gives the 'curvature component' of the geodetic effect.

74. A glimpse of something

To me, and perhaps to anyone who thinks PSG is right, this gives a tantalising glimpse of the refractive medium in action. A look at how the equation was arrived at shows why. First you put in a basic orbital speed across the whole object, taking its orbit to be circular. Gravity Probe B's orbit was deliberately made very near to circular.

I used a Newton circular orbital speed (a PSG speed can be used, which was found three years after this equation. That can be put in easily enough, but it reduces to Newton's theory anyway, and the difference is negligeable.) Next, the local slowing effect of the RM is put in to make slight adjustments to the speed of the object at two different points in it, which makes the upper edge travel faster than the centre.

The two places in the object are slowed by different factors, which leads to a difference in distance travelled over one orbit. From that the change of angle is derived via simple trigonometry. (The basic orbital speed isn't visible in the equation, as it gets cancelled by terms used for the orbit period.) What can then be seen is a basic orbital speed, and extra adjustments on top of it. And the point is, these two elements seem very separate - they're unconnected mathematically. So it looks as if something is making the object orbit, but *something else* is also affecting it, slowing the whole object down from that basic speed, but more at some points in it than others. This is exactly as PSG says it will be.

From my point of view, because I believe the picture I have is right - for good rational reasons - it was revealing that this gives the change of angle that we know happens, because it was measured. It meant I was seeing the refractive medium at work. There it was, doing exactly what I had believed it does for many years, and it was possible to trace its effect.

It fitted the effect in a detailed way: with PSG, if you make the object larger, but keeping it the same distance from the mass, the angle through which it turns will stay the same to a good accuracy. That's because the two edges, now further apart, will have a wider speed difference. So something general seems to be revealed. From my viewpoint the measurement showed the RM exists, and matter can be seen responding to this medium as light does. So although PSG wasn't the only explanation, that key area of the background picture, where matter behaves like light at a small scale, was also supported by the result.

As well as measuring the geodetic effect, Gravity Probe B also experienced some weird, unexplained effects. People were very much more interested in removing them as 'noise', that studying them as anomalies, or even as clues about the nature of gravity. But the 30 years of preparation the team did on it were to shield out all noise, leaving only signals from gravity itself.

So although experimenters are of course meant to find whatever they find - and not adjust it to fit expectation - they decided the unexplained behaviour was not part of the result they were trying to get, and removed it as 'noise' without knowing for sure what it was. They put it down to 'random patches of electrostatic potential fixed to the surface of each rotor', but this idea was not pinned down, as the word 'random' shows.

After the final results, a summary by the top relativist Clifford Will, quoted in Appendix C, says that there were 'extraneous torques', and 'some damping mechanism' at work. They found that an extra movement expected from the gyros, like the slow circular motion of a spinning top, unexpectedly slowed down over time, and the angle changed steadily. Will says the main angle also made large 'seemingly random "jumps" '. Although patterns were found within the effects, no-one knew what caused the effects, or the patterns.

This damping mechanism could perhaps be the RM, which varies in density slightly at different heights, affecting speeds differently. That behaves like a damping mechanism - it might or might not be the one in question. GP-B was a *very* sensitive system, and there's also the fact that, going by PSG, the slow geodetic rotation happens around a point very slightly below the geometrical centre. That wouldn't help. Only a mathematical analysis would reveal if one of these was the cause, and here they're just examples of possibilities.

But the data could contain proof of PSG. So some time in the future, if PSG is looking viable, requests for the data may be made, or funding raised for it be reanalysed by the Stanford team, NASA, or someone else.

At the time, the situation in gravity physics was more interesting to me than these anomalies. Just as an effect was finally measured that seemed to show space is curved, an alternative was published soon afterwards which *proved* mathematically that the cause could be something else. This widened the possibilities: now it could be either of two things.

So once again the RM picture had created an alternative interpretation, and had turned out to be capable of producing the same observations as GR, or mimicking them closely. But now it had extended the area it could do this in to matter as well as light. As a result, what the Gravity Probe B measurement showed was simply that what's known as the geodetic effect exists. It didn't prove - as many had thought - that space is curved. The question of whether space is curved or flat is still an open one, because we now have physics that works for either possibility.

Part 14. Strong gravity

75. Black holes

I was once in a bar where the conversation turned to black holes. I didn't say much, but I noticed that some took the name more literally than it should be taken. There are a lot of unanswered questions, but a starting point, to break this common misconception, is that what we call a black hole is a spherical chunk of matter. The name is a name. It's partly about how the object looks from the outside, and speculative ideas about the centre. But in terms of what we actually know, a black hole is not a hole - not literally.

A black hole certainly has an event horizon, or something very similar, which is a boundary beyond which nothing can escape. Whether a black hole has a *surface*, as a neutron star does, is an open question. There are arguments for and against, and evidence both ways. (People have argued that there can't be a surface, or explosions as on neutron stars would have been detected. But they admitted that in putting their argument together, they assumed GR is correct, and made other assumptions.)

We've studied black holes for seventy years, but for much of that time all we had was a pencil and paper. Or a typewriter, and by the late 20th century it was a computer. But it was still a case of assuming that general relativity is correct, and that in strong gravity, where a mass gets so compressed that nothing can escape, that absolutely no other effects come into play. And you assume this is so even if quantities become infinite, and even though general relativity itself breaks down.

This includes assuming that a number of things happen at the centre, such as matter becoming infinitely compressed, that are arguably very far-fetched in terms of what we know, and which may or may not be possible. Then you do some mathematics.

So for a long time, assumptions and calculations were all we had. Then near the end of the 20th century a little actual data started coming in. And by the 21st, new forms of astronomy were bringing in a flood of data, on what black holes are actually like. And when we were finally able to take a proper look, although it's too early to say for sure, what we started to find included some real differences from what we'd imagined. There were major surprises, but the public got the impression that everything was as expected.

At the extreme, where gravity gets very strong, our theories splay out. Even theories that have been equivalent in ordinary gravity, and impossible to tell apart, start to show real differences near the edge of a black hole. That's why people who study gravity find black holes so interesting, and why it's exciting that recent technology is producing an explosion of new data. We've hardly started, but there has already been a list of anomalies. These are not just the problems about the source of the LIGO data I've mentioned - some are about direct measurements. And although they might be nothing much more than the teething problems of some new forms of astronomy, what comes back is very specific, and it sometimes contradicts general relativity.

76. The background problem

Black holes have presented a difficult choice for physicists. Relativists tend to take the approach that general relativity can be used anywhere: they like the idea that there's nowhere we can't use it. In fact, it has only been shown to work over solar system distances, and has been unable to describe gravity at a large scale, for whatever reason. The jury is still out on the question of why, but one of the aims of putting in dark matter and dark energy (one of them is far more likely to exist than the other), is to have general relativity working well everywhere. I'll talk about those questions later.

With black holes, we either had to say we can't apply GR inside a black hole, because there are probably other effects at work - or we had to say that we can, but GR breaks down mathematically. The latter is what was chosen. It makes a black hole a little bit less of a no-go area, but at a very high cost. It means that in certain places our physics breaks down, along with its ability to describe the universe. So that also limits our ability to explain, get a grip on, and understand our world, perhaps into the far future.

On the other hand, many physicists think that when we eventually describe a black hole properly, using quantum theory and quantum gravity, there won't be any singularities. In one common view, better mathematics will remove a lot of the weirdness. So even though the work on black holes may well have to be thrown out later, there's still a reluctance to hold off, and say that we don't yet know how to describe these things. People understandably want to use GR as if it worked everywhere, but this reluctance to say 'we don't know yet' has been our downfall in many areas.

So physicists took onboard the idea of a singularity, and it became a part of GR. A few decades exploring the mathematics surrounding an idea makes us feel more comfortable with it. As we found out more on paper, loosely, some infinities turned out to be not at the edges, but at the centre. This helped, but they were still infinities. And at the centre, matter is thought to become infinitely compressed. In a common view of GR, *all the mass of the black hole is compressed to a point at the centre.*

But what we already know about matter makes this idea highly speculative. We know that matter at the quantum level is very like waves - it behaves like a wave, making interference patterns, just as real physical waves do. And yet matter is thought to become infinitely compressed, and still curve space to cause gravity, as in general relativity. But waves disappear if you compress them beyond a certain point, and infinity is a long way away. Somewhere before that, other changes would happen.

The idea that matter can be infinitely compressed, to me, is a good example of a 'mathematics only' approach. It's about treating matter as merely a mathematical quantity. But we know matter is more than that - it's real, and can show complex behaviour. What GR is really telling us is that it hasn't got enough information to calculate this. (Back in the '60s, sci-fi computers used to say '*that does not compute*', and I think GR is doing the same.)

So in my view there are some very questionable assumptions in applying GR to the interior of a black hole. The approach that would be more appropriate at present is to say 'so far this area of the picture is incomplete'. But instead, people have talked about a 'Golden age of black hole physics', and this was before there was even any direct data to compare with the mathematics. To me this shows what an unscientific attitude can look like.

The PSG picture is different. If PSG is right, it would remove something that might put non-negotiable limits onto our ability to understand and describe the universe, and has been a problem for both physics and philosophy. The PSG picture makes a black hole far less of a no go area, and would mean that sooner or later we'll probably be able to describe a black hole in a complete way, without any of our laws exploding (though I'm not saying we'll ever be able to actually enter a black hole without exploding). So PSG removes what might be a limit to the scope and range of science.

But it has to be said that the picture from PSG, although it makes important qualitative predictions in this area which seem to be appearing in the data in a number of places, is still very incomplete.

77. Black holes as in PSG

At the edge of a black hole, in either GR or PSG, the numbers go to the edge of - well, the numbers. At the Schwartzchild radius, some go to zero, some go to c, some go to infinity.

Escape velocity there is the speed of light. Matter can't reach c, so nothing can escape the pull. The escape velocity formula is the square root of $2GM/r$. Near a black hole $v_{esc} = c$, so for the radius where this happens, $r = 2GM/c^2$. Move that around a bit, and you get $2GM/rc^2 = 1$. And that, in turn, takes a very important number in the field (expression 4, p. 66), to zero. In either GR or PSG, near the edge of a black hole, $1 - (2GM/rc^2) = 1 - 1 = 0$. The square root of that is also zero, so there goes the time rate. It means if GR is right, time stops dead from certain viewpoints. This may or may not be possible in reality.

But in PSG everything is more literal. Expression 4, which goes to zero, is also the transmission speed of space. So if this is right, space no longer transmits waves at the Schwartzchild radius. So the question arises of whether that can happen. That depends on several things, but it becomes clear that it can't. If it could reach that point, this would also happen at the Schwartzchild radius, which is where everything happens - or stops happening.

As I've said, there's no knowing if other forces come into play when you get near the extreme. But even if you do to PSG the same thing that they do to GR, and assume that only the theory you're looking at applies (and ignore all other possible effects), you *still* find the extreme can't be reached. According to PSG, the secondary vibrations themselves make it impossible to reach the mathematical extreme.

Say a mass is accumulating matter steadily, with particles being added all the time. As its gravity gets stronger, it keeps emitting more and more secondary vibrations like a 'sparkler' firework. That's the refractive medium, it slows the transmission speed of space, and causes the gravity field. But when the field gets very strong, adding more secondary vibrations would have less and less effect. The secondaries slow everything, including themselves, and slowing themselves affects their ability to slow anything at all, and to do what they do. So you get a feedback loop. Near the edge of a black hole, where nothing can move easily, the secondary vibrations start to lose their ability to slow the transmission speed any further.

At the Schwartzchild radius, if the field could get to that mathematical point, nothing would be able to travel at all. Everything grinds to a halt. That's bad, but it gets worse - matter would also be unable to exist, because its nature depends on the fact that it rotates around the cylinder on which it lives. And gravity would stop working as well, because if matter stops rotating around the cylinders, it can't get refracted towards the mass, and there isn't even a pull trying to refract it towards the mass.

But none of that can happen. We know that waves can slow things on Earth, but only if they can *move*. If the secondaries can't even travel through that jammed up space, they can't do their job, which is slowing the transmission velocity, and so causing all these things to happen in the first place.

So a field could never reach the extreme point. As it approaches it, a small extra curve on the graph would have to be added to allow for this feedback loop, which begins to make a significant difference near the Schwartzchild radius. I managed to limit the possibilities about where this happens with a calculation, but a better model of the secondaries would be needed to do this in any complete way. If good confirmation for PSG is found, this is one of many areas where further progress might be made.

In this alternative picture, the secondary vibrations seep out slowly - slowly by a clock on Earth, for instance - from a huge centre of disordered vibration, where it's difficult for waves to travel. Moving away, they speed up steadily. Light also seeps out slowly from near the edge of the black hole. (In GR light does something similar, but due to the time rate.) In PSG the terrain it has to travel is different, and the effects there are more real.

Observable differences is the exciting place, affecting how a black hole will look to astronomers. And now we suddenly have all these new kinds of data coming in, we're going to find a few things out. The key prediction from PSG is that matter can and very often will be present, sitting right near the event horizon. According to GR there should be an event horizon of a kind we more or less understand, and any nearby matter will be swept inside it.

In PSG the event horizon is less sharply defined, as the extreme state can't be reached. The line will be slightly blurred, though how much is unspecified for now, and the difference might not be visible. This is not measurable anyway: astronomers have to remove some much larger blurring effects.

But there's something easier to measure. In PSG matter gets trapped there, right near the edge, and will be unable to move except very slowly, due to a medium with an enormous refractive index. That contradicts GR. In standard theory, matter inside a certain radius will be quickly swept beyond the event horizon. And so far, around five separate reasons to believe matter is there, near the edge of a black hole, right where it shouldn't be, have been found - and that includes many direct detections.

This could make a PSG black hole look fuzzier than a GR one, as light from near the edge has to come through any matter there. But apart from that, the object should look very like what a black hole is thought to look like. And recent observations have contained anomalies of several different kinds, that do indeed suggest matter is present. I'll set them out in a minute.

Standing back and returning to the overall picture, the idea of a black hole as like a stagnant area in a stream, where things have nearly stopped, leads to a rather different picture of the universe as a whole. At a large scale, you just have space and galaxies, in clusters and threads. We've known since the '90s that galaxies have huge black holes at their centres, though it was suspected before then. The stars and gas move around a central vortex.

But rather than being like giant 'plugholes', as in the present picture, in the view from PSG, galactic black holes look more like places where the 'flow' of the universe - of matter around the small-scale dimensions - has almost seized up. It's like a log jam in a river. The general blockage is caused by gravity, and it leads to huge local systems that slowly rotate, with intricate structure across many scales, consisting of beautifully patterned and ordered vibration. And occasionally - on at least one occasion and perhaps on others - with weird creatures living there as well.

78. Sagittarius A*

The idea of a black hole as in GR might be right. But as I've said, GR might be the truth but not the whole truth: there might be other effects near the limit. But we've assumed that GR is all that's needed and so far, in some areas, astronomy has backed this approach up quite well.

Objects like black holes certainly do exist. They've been 'photographed', and observed via radio astronomy, and we know beyond doubt that enormous masses can get compressed to very small radii. We don't yet know if they're exactly as in GR: they look similar, but there seem to be differences. There's one comparatively nearby, at the centre of our galaxy: Sagittarius A*, which has been closely watched since the beginning of the '90s.

That has included tracking the orbits of stars around it, and calculating the central mass from those orbits. The orbits agree approximately on the size of the central mass. It's small for a black hole at the centre of a galaxy, but it's huge compared to a stellar one. And it's clearly something very like what we mean by a black hole. And recently both radio astronomers and gamma ray astronomers (using long wavelengths and very short ones) have been getting direct data from black holes.

The giant black hole at the centre of our galaxy can't be seen with a normal telescope. To put it simply, there's a lot of stuff in the way. Radio astronomy, however, has come a long way since it started in the mid 20th century, and at longer wavelengths it has been possible to make out some structure there since 2007. But to do that, people have achieved something that would have been inconceivable a few decades ago, and have found ways to link up radio telescopes across the planet using computers, so creating the equivalent of a much larger radio telescope dish. Shep Doeleman, an American astronomer, is the man who has had the vision and drive to make this happen.

This is of great interest to astronomers and gravity theorists, because it has been more or less the first direct information on gravity at the extreme. And that's where it becomes easier to tell one theory from another, and to find out what gravity is really about. So the object at the centre of our galaxy may tell us a lot over the next few years.

We know its mass, assuming that our basic gravity equations still hold there. And they do seem to give roughly self-consistent results, when we watch the orbits of stars that travel around it. Estimates for the central mass vary, but it's somewhere around four million suns. This enormous mass is thought to be compressed into a sphere with a radius less than that of Mercury's orbit. Everything else orbits around it. A study of 6000 stars near Sgr A*, done over a few years, led to one of the mass estimates - the rapid motion of stars near the central mass make their orbits comparatively easy to pick out, from the much slower moving stars further out.

Some go so fast that they've been tracked through a complete orbit. So far only two have, but one in particular has been giving good data - it's a star called Source 2, or S2, and it orbits every 16 years. It has been followed since the mid '90s. S2 is a young star, and seems not to have had time to migrate inwards from further out. But it's not clear how it formed that close to the black hole, as a gas cloud would be pulled apart. This is one of a number of entirely unexplained things that have already appeared.

The Event Horizon Telescope (EHT) has been observing of Sgr A* since 2007, and increasing in resolution. Sgr A* is the only galactic black hole for which certain numbers are available, allowing anomalies to be found. It was imaged in 2017, with M87's black hole, and the images have since been shown to the world, without mentioning the problems. There'll be more news from the centre of the galaxy soon - until then, and without any doubt afterwards as well, there'll be a long list of unanswered questions.

79. The first gamma ray photos of black holes

PSG says that any black hole should have an area of 'choppy water' around it, where space is vibrating strongly at the very small scale where matter arises, so nothing can move easily there. There's likely to be matter trapped around the black hole, near the event horizon, or near where there's something that looks like one. GR says that any matter there will be swept across the event horizon, never to return. But a PSG black hole can have matter surrounding it, so it should look fuzzier: more blurred and spread out than a GR one. Light coming from behind the object has to travel through this area, and through any matter sitting near its edges.

And in the first gamma ray photos of them, black holes are indeed looking fuzzier and more spread out than expected. Naturally enough, other effects might be behind that, but it's interesting because there may be further clues to be found in the data.

Black holes can be seen at very short or very long wavelengths, but not much along the spectrum in between. In a 2010 NASA funded study, researchers made a composite image from 170 photos of the black holes at the centres of galaxies, which basically involved putting them all on top of each other.

Each image was made from very short wavelength, high energy gamma rays, taken from the orbiting Fermi telescope. The composite photo, with many AGN (active galactic nuclei) superimposed, was a way of picking out general patterns and trends in the characteristics of black holes, that a single image wouldn't necessarily reveal. The astronomers also made a computer model image - using standard theory, that is, GR - of what 170 black holes all sitting on top of each other should look like.

This is one of our first glimpses of black holes. When they put the real photo and the computer model image next to each other, they unexpectedly didn't match up. The real photo was a lot fuzzier and more spread out.

Because the composite photo is an averaging over many black holes, it might give some very general pointers about strong gravity. When it was found to be different from the GR prediction (like the radio astronomers who early on decided they must be looking at an orbiting hotspot, when the black hole was too small for what GR predicts for that size mass), the team who made the image of the black holes decided the cause must be something else. It was assumed that it showed the GR picture, but that it got distorted *later on*, while the photons were on their way to Earth. They came to the conclusion that a weak magnetic field was distorting the images.

GR is a well established idea, and people sometimes can't take an approach that's detached enough to see that it might be wrong. But a truly scientific approach includes that kind of detachment. The astronomers were so sure GR wasn't at fault that they announced the discovery of a weak magnetic field across the whole universe (I'm not saying they announced it across the whole universe, just that the magnetic field was thought to cover that area). Their paper said the weak magnetic field must be another relic left over from the big bang, like the cosmic microwave background.

But because we're now actually looking at black holes, instead of guessing what they're like, we might instead be seeing an unexpected aspect of them, that we simply didn't know about.

The magnetic field idea didn't hold up to careful scrutiny. A few months later another paper, from a different group of astronomers, showed that it wasn't a magnetic field causing it - they suggested it was an instrumental error. The first paper's title opened like this: '*Evidence for Gamma-ray Halos Around Active Galactic Nuclei...*', but the second paper's title rather pugnaciously began: '*No Evidence for Gamma-ray Halos Around Active Galactic Nuclei...*'. So as you can see, the result has been controversial - the cause is still an unanswered question, but some time later the suggested magnetic field had been more or less ruled out.

Gamma ray astronomy is a young field, and we don't yet know what it will find. But it looks like the blurring at those wavelengths will not be explained away, and it may strengthen other evidence for the presence of matter. The result may not leave general relativity looking as good as it did.

There are other theories that have no sharp event horizon, it's not just PSG. But with PSG this point is hard to pin down, because the amount of matter present affects the blurring. And astronomers doing measurements have to try to remove many effects anyway.

But if the presence of matter is what we keep finding when more data comes in, GR can't explain that. The real cause might be that a gravity field consists of vibrations emitted from the mass. At the extreme they capture matter, leading to an object that looks different. Electromagnetic radiation from the black hole might be more diffuse and spread out than GR predicts. If so, what we saw in that composite photo, averaging out 170 black holes, is what a black hole actually looks like.

It's worth pointing out that early attempts to study black holes, via different forms of astronomy, have led to anomalies at different wavelengths: radio astronomers found anomalies at long wavelengths, gamma-ray astronomers, with gravitational wave astronomers found (in the optical counterpart of a LIGO detection) an anomaly at very short wavelengths, with the presence of matter. And there's also the gamma-ray composite photo.

The chapter after next lists these anomalies, including a fourth one, involving echoes found in LIGO's gravitational wave measurements. It's very drastic in its effect - if confirmed, it would falsify GR. A further anomaly was a 68 solar mass black hole, which according to standard theory shouldn't be above 25 solar masses. These bits of evidence, perhaps taken together, may eventually disprove general relativity. But at present it's too early to tell.

80. The orbit of S2

Somewhere before the Schwartzchild radius, as one approaches a black hole, according to PSG there should be a 'saturation point', where gravity can't get any stronger. In GR the Schwartzchild radius is the limit, but in PSG there's a limit before that, slightly further out, where the secondary vibrations start to weaken their own effect significantly. And at this saturation radius, matter can only travel slowly (by a distant clock), and things might get very blocked up, but everything can still travel.

It's not far from the Schwartzchild radius. The full PSG orbit speed equation, post-Newtonian adjustments included (it's explained in Chapter 99), is:

$$v = \sqrt{GM\,(2/r - 1/a + [2GM/arc^2])\,(1 - [2GM/rc^2])} \qquad (42)$$

It gives results very like standard theory for the orbit of S2, which has been tracked around a whole orbit, leading to a mass value of 4.1 million suns. The astronomers published a simple Keplerian orbit, and apart from a very slight difference to the central mass value, the two orbits look the same. It's 'a very slight difference', but though negligible in most ways of looking at it, it's nine thousand Earths. That just reflects the scale: a few million suns.

S2's closest approach is 18 billion kilometers from the black hole, and GR and PSG are still roughly the same there. But at the Schwartzchild radius, only 17 light hours away, the two theories have diverged, and are very different. The above equation goes to v = 0, if applied as it is, but going by standard theory v = 0.7c. At the Schwartzchild radius these are both hypothetical, but they're not hypothetical a little further out, and the two theories disagree on what happens in the region right near the black hole. And so far the data has been suggesting PSG there, in two separate ways.

We'll look at the accretion disk in the section about predictions. But briefly, GR says matter has no stable orbit inside three Schwartzchild radii, so it falls in immediately, and can't stay there. PSG says matter *can* stay in place inside this limit (matter has been found there, beyond any doubt), and that outside it, orbit speeds will be significantly slower than Keplerian ones. Both of these predictions seem to be turning up in the data.

So the two theories give different sets of numbers for the orbit speeds. One great thing about black holes - they make theories distinguishable from each other. It may be possible to match orbit speeds at different radii to the PSG orbit speed equation (the simple circular orbit equation is in Part 18).

There's another star near the black hole, called S0-102, with an even shorter orbit, of around 11 years. It's fainter, but it has now been tracked through a complete orbit as well, and the two stars together should provide some good ways to test GR. In 2018, S2 passed its closest approach to the central mass, and the astronomy team, led by Andrea Ghez, who have been studying the region since the '90s from the summit of Mauna Kea in Hawaii, measured the wavelength of the light from S2 (the gravitational redshift), showing how the light was stretched by the surrounding gravity field.

The gravitational redshift is exactly the same in PSG as in GR, so the fact that it came out right doesn't tell us much on that particular question. But there are other differences - anywhere else they'd be hard to measure, but at the extreme, some differences surface. Both stars are on quite eccentric elliptical orbits, which helps. And there are two, which helps with ruling out sources of error. So there's a real possibility of looking at the orbits, and seeing which theory fits the orbit data better.

81. The strongest evidence against GR, from new technology

This chapter has a list of some of the strongest reasons to think GR is wrong. People say GR has 'passed every test with flying colours', but the full impact of this statement comes from leaving out the tests it has failed.

Stacy McGaugh, Professor and Chair of Astronomy at Case Western Reserve University Ohio, puts it better. In an excellent interview, he was asked about a truly baffling puzzle: synchronised satellite galaxies - the 'planes of satellite galaxies problem'. It surfaced only recently, in 2018.

Larger galaxies very often have small satellite galaxies orbiting them. The GR + dark matter standard view (λCDM), says they should orbit randomly. What they do instead is orbit in the same thin plane, most of them travelling in the same direction. In standard theory, they shouldn't do this, except in 0.2% of these situations, just by chance. But instead they do it a lot: the first three of these satellite systems we've looked at all do it.

This goes against the standard Lambda cold dark matter λCDM view, which is very detailed, and should explain anything of this kind. Not only does λCDM fail to explain it, so does *everything else we have*. All dark matter theories, all gravity theories. And being a problem for dark matter, this is also a problem for GR, as dark matter was designed to rescue GR where it fails. I'll set out a possible solution for this puzzle later. Asked how serious a challenge to the standard view it is, Stacy said:

"Serious" depends on perspective. If you are utterly convinced that the current cosmological paradigm has to be correct, then this new body of evidence is a minor detail to be brushed aside, with the expectedly mundane explanation worthy only of debate by hyper-specialists. If you have serious doubts, this is another in a long line of evidence suggesting that something is screwy.

Stacy McGaugh is an excellent astronomer and physicist. That interview (see the reference section) is well worth reading, and it also deals with the 'cusp-core problem' with dark matter halos. Stacy has pointed out some important flaws in the present picture, and shows that it doesn't fit together. Although I ended up reaching a different view on the mass discrepancy, his work was a large part of what led me to it. But it's not a case of disagreeing with him, as he doesn't suggest causes. Instead he points out vital clues - mathematical connections - that have been overlooked. His interest in MOND means he may see GR as incomplete rather than wrong. 'Something is screwy' implies *something*, somewhere: in dark matter, or GR in the wider domains, or both. We'll come back to Stacy's work later, with the mass discrepancy.

But talking of the mass discrepancy, I'll mention something else briefly. It's not only dark matter that sometimes looks bad, it's GR as well. The best 'map of dark matter' so far was published in 2021, by DES, the Dark Energy Survey. It suggests that general relativity is wrong. To understand how physicists feel about the situation, Niall Jeffrey, who put the map together, was quoted as saying that the result poses a 'real problem' for physics. *"If this disparity is true then maybe Einstein was wrong. You might think this is a bad thing, that maybe physics is broken. But to a physicist, it is extremely exciting."*

Professor Carlos Frenk, whose contribution to the present cosmology theory, λCDM, was a major one (using general relativity) had strong feelings about it: *"I spent my life working on this theory [GR], and my heart tells me I don't want to see it collapse. But my brain tells me the measurements were correct [...]. Then my stomach cringes, because we have no solid grounds to explore because we have no theory of physics to guide us. It makes me very nervous and fearful, because we are entering a completely unknown domain and who knows what we are going to find."*

Among the many other points that suggest GR is wrong, there's the 'Vacuum catastrophe', which is outlined on p12-13. It's described in detail in Book III, because to give my solution, you need more of the picture from underneath the background theory.

I'll list some of the black hole anomalies now, most of which can potentially be explained by matter sitting just outside the event horizon. There's LIGO's first gravitational wave detection: we observed a merging of two black holes, with a burst of gamma rays immediately afterwards. The puzzle is that the gamma rays mean matter was present, getting smashed up in the collision. GR says this is impossible, as it would have been swept up before then.

The PSG picture can explain the gamma ray burst. It's from matter that gets stuck in the strongly vibrating space, with a huge refractive index. The odds that the detections were separate events, coincidentally timed and aligned, were calculated to be 1 in 450. So it looks like the same event, and from a GR point of view, the measurements are telling us two conflicting stories about that event.

Then there's the unexpected fuzziness of the composite photo of many black holes, which may be caused by the same thing: matter where it shouldn't be. Another anomaly is about the inner region of Sgr A*, our galaxy's black hole. GR says the accretion disk, a swirling circular disk of matter being pulled in, has an abrupt cut off point, at 3 Schwartzchild radii. Inside that radius, there are no stable orbits, so any matter will very quickly fall in. Light can orbit at 1.5 Schwartzchild radii, but not matter. So there should be an outer disk of matter, with a circular gap near the centre, like a vinyl record.

But matter has been detected, beyond doubt, well inside the cutoff point. So the standard picture is simply looking wrong: people have come up with all kinds of explanations. They talk about a 'viscosity' near the black hole, which is a new way of describing spacetime, as if it has the properties of a medium. It becomes particularly relevant near a black hole. In PSG it's a highly dense refractive medium. There's more detail on this in the predictions section: the PSG equations give noticeably slower orbit speeds in the accretion disk than standard theory. Speeds there have already been measured - they're slower than they should be. But other effects need to be taken into account, and it's not straightforward.

But recently, something else has come up, and yet again, it might be about matter near the event horizon. It actually contradicts GR in a direct, hard-to-refute way, and several physicists have said in interviews that it means we may need to rethink gravity. Some researchers predicted 'echoes' that would be found in the LIGO gravity wave data. They were searching for something rather different from what they may have found.

They were looking for a 'firewall', which is like a ring of high energy particles around the black hole. Instead of the GR type event horizon, there would be something else - a spherical boundary, relating to quantum mechanics, but inconsistent with GR. These echoes should follow the main LIGO signal soon after it, in the form of a smaller signal. Whatever the cause, it would mean the gravitational waves were bouncing around inside a spherical boundary, and some of them took a longer route before getting out.

If they found the echoes in the data, it would mean there was something to bounce off there. This spherical barrier could be a firewall, as the researchers thought. But it might again be matter around the black hole. The idea of a firewall is arrived at via complicated and far-fetched reasoning, and it isn't necessarily even viable: the group who coined the term 'firewall', identified what they called the 'firewall paradox'. If a firewall existed, infalling matter wouldn't get past it. This would stop the build up of mass we that know for sure happens in a black hole. So instead of a firewall, it might be matter near the horizon. But *either way, it goes against GR.*

The group that searched for evidence predicted the pattern, looked for it in the data, then found it. It was potentially strong evidence against GR, and if confirmed over many results, it would quite simply falsify GR. So it seems the gravitational waves LIGO detected were bouncing around before departing, and according to GR, they just wouldn't do that. The people studying it then found similar patterns in data from other black hole mergers, which means the discovery is probably real.

But others questioned the findings, or tried to put a different interpretation onto the data. Then many started searching for more 'echoes', and in 2022 a highly respected team from the Perimeter Institute published a paper with *very* strong evidence for a newly discovered echo, and in the largest event of this kind ever found, which makes it much easier to identify the echo. So the latest measurements have now made the discovery unavoidable. If so, GR is wrong. But resistance to evidence against GR is very strong, in some areas to the point of bias. The investigation is ongoing, and it's too early to draw final conclusions - further data will shed more light on the question.

But in general, if these anomalies are caused in the ways I've suggested, it'll be increasingly apparent. If this approach is correct, with these new forms of astronomy just taking off, in the future the differences will be there for all to see. Whatever really goes on at the edge of a black hole, the evidence is out there, it's all over the universe - which makes it hard to sweep under the carpet - and it's not going to go away.

Additional note:

The last few sections of the book, from Part 17 onwards, are what I've called an in-depth area - it's more for people who have studied physics. But the last section, Part 21, sums things up, and is for anyone who finds it of interest.

Before the in-depth area, the next two sections are about two puzzles for which PSG has no complete explanation: the flyby anomaly and the mass discrepancy. So some areas of Parts 15 and 16 are more speculative. But they're both absolutely crucial puzzles, and an initial solution can be qualitative rather than quantitative, as in the points I've made on black holes, which arise from PSG. And in the case of the mass discrepancy (the dark matter puzzle), what I'd say is a very much needed solution, arising directly from PSG, is put forward.

The flyby anomaly is the only area where PSG has no solution. But as I hope and expect some readers will find, aspects of that puzzle are hard to understand in the context of GR, but far, far easier to understand via PSG.

In the in-depth area, Part 18, 'More mathematics: orbit corrections' in particular contains central, hard and fast parts of the theory.

Every theory has to be testable, otherwise it's only a hypothesis. The in-depth area includes experiments to test PSG, set out in mathematical detail. To me this is one of the most exciting parts of the book, because (and this is the great thing about physics!), it's about what can actually be found out, and established for certain.

Part 15. The flyby anomaly - a major puzzle

82. A new mystery arrives

There's an important aspect of solar system gravity that I've not mentioned, and here we come to another unsolved mystery. So I'll go through the clues as always, but in places it'll be more as a fellow mystery enthusiast than as someone who's putting forward a solution. That will make a change, and not having an answer actually frees me up at times, to wonder with less pressure about the clues, and what they're telling us.

After many years of work on it I still have no full solution, but to me studying the flyby anomaly is interesting in itself. It also makes a good story, involving the interaction of a mixed bunch of physicists, and shows how science works when it gets into action, with a problem to fix that can't be ignored. Although I haven't explained the flyby anomaly in any complete way, it nevertheless points at PSG - but the story is worthwhile anyway.

Way back in the early '90s, some NASA people started to notice slight speed discrepancies when space probes flew close by the Earth. Differences would be found between the expected speed for the probe, and the measured one. Every few years NASA would fly a probe near the Earth for a 'gravity assist' - a boost to its speed, gained by swinging close by a planet and off away again. The probe curves around the planet, and is sent shooting off at a new angle. Jupiter's large mass is good for sending probes to the outer solar system in this way, but sometimes a probe has to jump around the inner solar system like a piece on a draughts board first, perhaps off to Venus and back, just to get to its real destination.

It might seem that the planet's gravity is what makes the 'slingshot effect' of a gravity assist. In fact on its own the field does nothing: it speeds an object up on the way in, and slows it down on the way out, with a total change of zero. What does make a difference is the planet's motion around the Sun. Jupiter's large mass moves along with plenty of momentum, and that's what makes it good for catapulting our little robots out and away. This has been done with different planets for decades - the only difference when the Earth is used is that our knowledge of the gravity field is far, far better, so anything odd might be noticed.

And something was. There were slight speed differences from the calculated trajectories on quite a few of the flybys. It was often an increase in speed, and no clear explanation was found that fitted the data. But instead, some strange patterns in the data were found, and it was these patterns that made the puzzle so interesting.

The flybys turned out to be an experiment to test gravity that no-one had thought of, because the probes are tracked with great accuracy. If someone had suggested it as an experiment, it would have been a very good idea, but nobody did. Instead we stumbled onto it while trying to do something else. This happens a lot in science - many discoveries have been made while trying to do something entirely different.

Anyway, the flybys led to a major mystery. For years it was overshadowed by the Pioneer anomaly, which seemed the larger puzzle of the two. But when that was solved, the flyby anomaly emerged as one of the greatest mysteries in gravity physics. Its importance is not to be underestimated - it's not a side issue. The problem is enormously important in physics today, partly because our view of gravity is constantly under question because of the long standing failure to find dark matter, and other cosmological problems.

Unlike with the Pioneer mystery, not many are saying that the flyby anomaly might have some everyday cause. The Pioneer anomaly turned out to be just heat escaping unevenly, and bouncing off the radio dish, pushing the probe slightly backwards, all the way across the lonely darkness of the outer solar system. What can you do? Don't say it's a tinpot contraption, it's just that it was designed and built in the late '60s and early '70s, and considering that, Pioneer 10 achieved great things.

People had been suggesting everyday explanations for the Pioneer anomaly throughout, along with some much weirder ideas. The number involved - the consistent, small sunward acceleration - landed near two important numbers in physics and cosmology, which kept people studying it for years. I was one of them, though I thought that the similarity to the other two numbers was coincidence. Most of the effects from PSG were far too small to cause it, but to me it looked like a discrepancy between two different hyperbolic orbits, the expected one and the measured one, as they got further and further out of sync due to slight differences to some of the parameters. But the number I hoped to see coming out of the calculations never did.

Anyway, some thought the Pioneer anomaly might have an everyday cause, and that turned out to be right. But with the flyby anomaly, it's much harder to think of anything we know about that fits the clues - it's at the very least something NASA haven't taken into account. But it might be new physics, and that includes the ring of dark matter surrounding the Earth, which has been one of a few suggestions. But what many of us have been wondering about, looking closely at the clues, is new gravity physics.

83. What happened in 2008

In 2008 the NASA team led by John Anderson published a paper that showed a pattern in the data. Until then all we had was a set of unconnected speed discrepancies for six flybys. The numbers varied quite a lot, and at first they seemed to be random. At closest approach, the speed increases ranged from around 0.01 mm/sec to 7 mm/sec. Given the precision of the tracking, that's a large anomaly.

Anderson had been the discoverer of the other puzzle, the Pioneer anomaly, decades earlier - he studied it through the '80s without telling anyone, which is understandable. Anyone who finds something like that always wonders if they've made an error somewhere. He eventually told Michael Martin Nieto, who says when he heard about it he 'nearly fell off his chair'. Later Anderson headed NASA's top team on solar system gravity, and for years was the co-ordinator of efforts to solve both puzzles. One day a member of his team, James Jordan, was trying things with the flyby data and looking for clues, and he found a thread that pulled all the numbers together.

A common way of seeing the anomaly is a sudden speed increase at closest approach, as the probes swung by the Earth. But in fact, there's an incoming arc and an outgoing arc, and the two arcs are so different that they can't be described as the same orbit at all. The Doppler and ranging data are fitted to a Newtonian hyperbolic orbit, which is a curving path. And oddly, we were getting two different hyperbolic orbits, one for the incoming path, the other for the outgoing one.

This actually rules out a lot of possible solutions. If gravity is simply different from how we imagine it in one way or another, it still might make hyperbolic orbits curve inwards, and then off outwards in just the same way, implying that the field is self-consistent. So what one might expect to find is the same anomaly on both the incoming and the outgoing arcs - but producing data that's difficult to fit to a Newtonian (or GR) orbit. But instead there was this difference between the arcs, and it seemed to be telling us that something was making the situation asymmetrical.

What James Jordan noticed was also about the symmetry of the situation. It was about the symmetry of the flightpath angle. The first thing he noticed - I'm guessing his route through the puzzle, but it seems very likely - was that one particular flyby created no anomaly. But there was something else about it as well. It also had almost exactly the same incoming latitude and outgoing latitude. That's about the angle to the equator: the flyby that created more or less no anomaly came in at roughly the same angle to the equator as the angle at which it went out.

He then found that by contrast, the flyby with the biggest anomaly had the *most* asymmetry between the incoming and outgoing angles to the equator. So Jordan started looking for a formula, and after a while he found one that fitted all the flybys very neatly - the ones that had happened up to that time. What he found was an empirical formula, which means that it fits the data, but you don't know why. But the formula was really 'semi-pirical' (as in the shorthand of my notes), because it gives you some idea.

The formula suggested that it's to do with the Earth's rotation. The direction of the rotation is along the equator line, and the formula contained angles to the equator. Now according to GR, the Earth's rotation can simply be ignored (although it creates other far smaller effects). If it *can* be ignored, as GR says, then all hyperbolic orbits will behave as if they're symmetrical, as if the Earth isn't rotating. But if the Earth's rotation is important, which goes against GR, some hyperbolic orbits will not be symmetrical after all. The Jordan formula contained not only the incoming and outgoing angles to the equator, it also included the Earth's angular momentum and average radius.

And those last two terms multiplied together give you, quite accurately, the rotation speed of the Earth at the equator. So the clues were pointing at the planet's rotation. And when a probe swings past a mass, the rotation of the mass is one of the few things that might remove the symmetry.

When the Jordan formula was published by the Anderson team in 2008, it sent ripples of excitement through the online physics community. The new challenge, instead of explaining the anomaly, was to explain the formula. So things had got more specific - people waited eagerly for the next flybys that were planned, to see if the formula held up. There have been several flybys since then, and surprisingly, none have seemed to produce any anomaly at all. The puzzle suddenly seemed to have completely disappeared, and just when we thought we were getting a handle on it.

Anyway, over several years, a German physicist called Busack is said to have predicted the null results for the last few flybys, as well as the earlier results, using a different approach. His work is so complicated that few have tried to check it - it involved a massive computer analysis of the whole situation. His idea is empirical: he doesn't claim to know what causes it. But he has found the numbers fit, to whatever extent, if he assumes the Earth's gravitational field takes on an extra asymmetrical aspect because of its motion through the cosmic microwave background, in the direction of that motion. Other physicists have published attempts to explain the anomaly as a variation on inertia, a local ring of dark matter, and a range of other things.

But one thing I noticed, and Anderson's team have mentioned it recently as well, was that the flybys that gave null results didn't come so close to the Earth as earlier ones. So perhaps the effect is a short range one - this would make the Jordan formula incomplete. But it really must be anyway, as it gives no range limit, and no way for the effect to decrease with distance.

The Juno probe flyby in 2013 was awaited by everyone studying the anomaly with great excitement. It went quite close to the Earth, so the Jordan formula should work. The Anderson team had accordingly made a prediction of an anomaly of + 7 mm/sec, and had been preparing to announce the result at a conference after the flyby, confidently submitting an abstract a little before the flyby, saying that they would talk about whatever they measured when it happened. The paper was called *'Juno Earth Flyby as a Sensitive Detector of Anomalous Orbital-Energy Changes'* .

Busack had predicted - 6 mm/sec (minus, that is), a very different prediction, and both were easily measurable. So it looked like the Juno flyby was going to decide once and for all between the two leading theories on it. The flyby happened in October 2013. At closest approach the probe was eclipsed by the Earth, and went into safe mode, due to the need to save battery power. So it didn't do much for a while, but the flyby otherwise went well.

After the flyby there were rumours of a null result, then of an anomaly, but the rumours didn't go into mm/sec. The NASA team then pulled out of the conference, the conference organisers said it might be due to health issues, and everything has gone very quiet since then. No announcement has been made about the measurements, and it has been years now. That's not the normal pattern, there's usually an announcement sooner than that.

In terms of the result, this could mean anything, except the Jordan formula having been correct. If it had been, we'd know by now. If Busack's prediction was right, which doesn't seem likely, then technically the relativity principle would have been falsified by prediction and experiment, and Busack mildly points out that his work goes against it. But he also says he doesn't know what's going on underneath his formula, which means the relativity principle (which I personally believe in, along with the NASA team), is comparatively safe. It's more likely that new clues have appeared in one form or another, or perhaps simply a hard-to-interpret null result.

84. The wilderness years

We rather lost the thread of the puzzle for a few years after the Juno flyby. This chapter is short, because the main thing about the wilderness years was that nothing happened. Anderson had reached retirement age quite a while earlier, but had gamely kept on working on the anomaly. But now we heard nothing more. The Jordan formula had failed to predict the null results from the last few flybys, so our handle on it had gone. But the formula can still hold clues, even if it's wrong, incomplete, or an approximation. Then a Dutch team published a paper on the anomaly on the last day of 2015, concluding that solar radiation pressure might have caused it. They settled one mystery, saying there was no recorded anomaly on the Juno flyby, which they knew from email exchanges with the Anderson team.

The man who stepped in and effectively took over from Anderson and his team, and became the leading investigator into the anomaly, was Luis Acedo, from the University of Valencia in Spain. He started alone, then collaborated with another physicist, then gathered a team around him. And when Juno made it out to Jupiter three years after its Earth flyby, in 2016, they started comparing its highly eccentric elliptical orbit around the giant planet with a very accurate model.

They also re-analysed earlier NASA data from Earth flybys. So now we might settle a few questions - is the anomaly universal, or just about Earth? Will it be larger because everything about Jupiter is - its mass, rotation speed and so on?

And then the anomaly finally reappeared. And it was indeed larger out there, a lot larger. This fitted with the Anderson team's view that it's to do with the planet's rotation speed and radius. So it looked like our handle on it, though certainly incomplete, had nevertheless been partly right.

85. The flyby puzzle: clues and possibilities I

I should set out some of the clues, and there are now new, quite weird ones, coming from Acedo's team. One earlier clue I've mentioned, which led James Jordan towards his formula, is like the well known clue of the dog that didn't bark - it's about something that didn't happen, rather than something that did. When the probe's flyby path is symmetrical to the Earth's spin axis, and so also to the equator, you get no anomaly.

My interpretation for that took years to arrive. It's that we're seeing part of an effect, not all of it. All our numbers for the Earth's gravity field, including for the extra mass of its 'midriff bulge', come from studying the gravity field of a rotating planet. We can't stop the Earth from spinning.

And our numbers, which we take to be the basic ones, come from orbits that are symmetrical to the equator, such as nearly circular elliptical orbits. We've built up our picture of gravity that way - studying symmetrical orbits around a rotating mass. So if there's an effect to do with the planet's rotation, part of it might be already built into our numbers - the part that shows up on symmetrical orbits. But the part that shows up on asymmetrical orbits might be what we call the flyby anomaly.

The only orbits that get very asymmetrical are hyperbolic orbits, such as the Earth flybys, and highly eccentric elliptical orbits, like Juno's around Jupiter. Before we started making probes fly past the Earth, we hadn't really messed with those kind of orbits. (Well, we had with comets, but we didn't have all the numbers.) So the flyby puzzle might be about what happens when we finally start to look closely at asymmetrical orbits, and find out that our view of gravity needs to be adjusted for them.

But this also means we might be looking at an incomplete picture, and seeing only part of an effect. The other part could already be in our numbers, or our equations. The clues show the anomaly doing more than one thing - it does several. But what we call 'the anomaly' is simply the *difference* between our present expectation of the behaviour of gravity, and the observed behaviour. So if our present numbers are hiding part of the effect, because it's already included in them, that's not going to help.

That may be why the mystery is so hard to solve. I've tried for years, but I've not been able to find a solution that can be pinned down. Still, I have an idea of what's going on. I'm not necessarily giving up working on this puzzle, but I've done too much time on it. I hope others will come in and try things with the angles that I'll mention now, or solve it from some different perspective. If they get there from the clues I'll give here, they may prove PSG right. The data is very accurate, and it might suddenly fit everywhere. I'll also send the ideas to Luis Acedo, in the hope that he'll look at his model again in the light of this new approach.

As I mentioned before, the background theory went into too many different areas to cover them all. And as always, it's a team effort, and many heads are better than one. Anyone trying to solve the flyby anomaly can certainly write to me, and if I can help in any way, I will.

86. The flyby puzzle: clues and possibilities II

Although it's incomplete, I'll set out the picture I have for the flyby mystery, which I'm sure some people reading this will have anticipated. Imagine the probe steadily approaching the Earth. It starts to plough its way through the medium made of small-scale waves, which gets increasingly dense. When the probe arrives near the planet's surface, the medium is pouring out there.

Now if the mass was sitting there *not* rotating, it's very simple. The medium is being emitted in all directions radially. The only irregularity in the outward moving pattern is anywhere the mass isn't quite spherical.

But if the mass is rotating, it's different. You might then get a swirling pattern in the medium. The nearer the equator, the faster the surface of the planet is moving, as the medium is emitted from it. At the poles there's only rotation 'on a sixpence', but travelling South from the North pole, the rotation speed increases, and it depends on the latitude.

Now suppose the planet's rotation makes it emit the medium at an angle, or with an additional sideways speed component, so that after being emitted, the medium moves around with the mass, at least a little. It might only move with the Earth very near it, but if so, this will happen differently at different latitudes, because these places are rotating at different speeds. The resulting 'swirled' pattern would be built into the medium as it emerges.

And yes, it could make flybys do odd things, particularly ones where the path is asymmetrical to the equator. This scenario is also more likely because the Earth bulges around the equator (as a result of its rotation over time), so the mass is uneven as well.

And interestingly, of the various explanations that have been put forward to interpret the mathematics of the anomaly, there's one that has been taken very seriously, called retardation. There are two separate theories, which the authors point out are different, both of which involve adding in a lightspeed delay to the effect of gravity.

But people have struggled with this issue before. They have for hundreds of years. Does gravity take time to have an effect? No, and if it did we wouldn't be arguing about it, because the solar system would have quickly collapsed. And that's not only in Newton's theory - the mathematical quirk in GR also means that in reality, gravity acts instantaneously. So when Joseph Hafele (decades after the well known Hafele-Keating 'jetlagged clocks' experiment), added a time delay into the equations, as if the effect of gravity travelled at the speed of light, no-one took his idea seriously - not at first. But he found that something very like the flyby anomaly came out.

And a physicist named Bel, who published with Acedo, also found assuming retardation of gravity led to something like the anomaly. They emphasised that the theory was different from Hafele's. Perhaps you can see how this point supports PSG, with its emitted outward-moving medium, at lightspeed. The point is, *something* seems to be travelling at lightspeed, and creating a lightspeed-type delay. But we know it's not the direct effect of gravity itself, because if it was, the solar system wouldn't be here.

So something is needed that *travels at lightspeed, isn't gravity, but is closely associated with gravity*. That looks just like the refractive medium from PSG, and without it, it's very hard to explain. According to PSG, the instant aspect of gravity comes from the refractive medium, which is already there to affect the object in question, out in space. This is not connected with its emission: gravity doesn't care how the medium travelled to get there.

The result is instant gravity, just as we've found to be the case. When gravity pulls on something, the cause is a local 'density slope' - but the medium has nevertheless travelled from the mass to the object it affects, and takes time to reach it. And that process of emission and travel may create a separate effect, at close range. Perhaps if the mass is rotating, the direction of the pull is not an exactly radial line, or the delay due to travel time affects where the mass appears to be in relation to gravity - in one way or another. From a PSG perspective, the clues suggest something of that kind.

Someone at NASA called the flyby anomaly 'super frame dragging', because although it's far larger than the frame dragging effect from GR, it's similar in what it does. So there might be a dragging of the medium up very close to a rotating mass. Luis Acedo, in 2015, looked at the idea that the anomaly was due to 'a strong transversal component of the gravitomagnetic field'. But by 2017 he was suggesting it arises from 'an unknown fifth force with latitude dependence', with a range of around 300 km. He says: 'A significant radial component was found, and this decays with the distance to the center of Jupiter as expected from an unknown physical interaction'. It relates to the centre, and that's a clue. It rules out a lot of non-gravitational effects.

87. Oumuamua

The Juno probe showed that the flyby anomaly affects other planets apart from Earth, and can happen on elliptical orbits, as well as hyperbolic ones. It may also affect objects that come from outside the solar system, well, one at least. An object that was named Oumuamua (by Earthlings when it got here), arrived from outside the solar system in 2017. It passed close to the Sun, and when it passed by, it seems to have had an unexplained boost to its speed. It went off away from the Sun at a different angle, and on a different trajectory from the expected one. This led to a lot of speculation about it being an alien spacecraft, and even from a few respected physicists.

And yet Oumuamua had done something not totally dissimilar from what our own probes do when they fly close to a mass. It had an unexplained boost to its speed. I'm not saying Oumuamua was therefore one of ours, I'm saying it was therefore a chunk of rock. Invoking alien technology to explain the speed boost is more far-fetched if we admit, as NASA do, that we've got a speed boost anomaly around here already. It's one that affects hyperbolic orbits, like that of Oumuamua.

But some are very certain that GR is right and complete. Speculation about alien technology usually comes from non-scientists, but at times it may come from physicists who are in denial of the utterly reasonable point that there could be something wrong or incomplete in our view of gravity. I don't know if they compared the data with flyby anomaly data before considering aliens, but there was a need to. The effect seemed different from the flyby anomaly in some ways, but it's an unsolved mystery, and the flyby anomaly has shown us a range of effects we don't understand (Section 17), not just one.

And Oumuamua was only noticed *after* it had passed close to the Sun. All our ideas about its earlier speed boost were guesses. In the light of that, claiming a reliable (30σ) measurement of a 'non-gravitational acceleration' is wrong. The measurement was reliable, but of an unexplained acceleration.

If these effects stem from a single cause, it starts to look like a universal one, and one related to gravity. Either way, an open mind is needed. We need to accept a fact that many have faced up to: that the long search for a genuine, complete understanding of gravity is quite simply not over.

NB. There were three more chapters at the end of this section, partly about the discoveries Acedo made about the flyby anomaly, and the strange clues he and his team found, with the Juno probe out at Jupiter. They include more ideas towards a solution, and start with what may be the best clue we have on the anomaly. Those chapters have been moved to the 'in-depth' part of the book, near the end - they open that part of it, with Section 17. But for now I've decided to leave the flyby anomaly to one side, as we should get straight to Section 16, and the mass discrepancy, which is a very pressing question.

Part 16. The largest puzzle of all?

88. Cosmology

Cosmology is like finding yourself in a forest, where the trees change slowly over time, and trying work out how they change. They're on a different time scale from yours, so all you have is a snapshot of one moment in a long, slow process.

I had a cosmological theory until about 2002, then decided to abandon work on that whole area. It had ideas in it that could be right - it's possible. But I don't know. It had mathematics and conceptual ideas, but there was no way to get to grips with it, and I found myself realising that I wasn't sure if it was on the right track, and there was no way to know. And by then other parts of the theory were shaping up well, and there were already some areas that I was absolutely certain were right.

It's no good just throwing the best guess you have at something. People do that sometimes, but the truth is, in physics, if you don't know something you often can't guess it. Some don't realise it, but the conceptual side is just like the mathematical side in that way. In either of the two sides, there's a lot to choose from. So if you don't know enough about what you're looking for, it's like a needle in a haystack. Because of this, it helps to admit what you don't know. At the time I had do that, and to choose what to work on, and I chose a set of ideas that was far more certain.

When I let go of cosmology, I felt there were too many variables (conceptual variables), and too many different kinds of things that might be going on, for me to apply my particular method for probing physics to that field.

If the picture of the current situation in cosmology that I'll present now looks less than optimistic, I'm sorry. Some have a too comfortable picture in which everything is nearly tied up, and we'll find dark matter soon, which is the last piece of the puzzle. But they thought that about physics at the end of the 19th century, just before it turned out to be entirely different, and the most baffling puzzles ever to arrive on our doorstep arrived on it.

Incidentally, in 2022 I found a new, entirely different cosmological theory. It came from an idea that went back 15 years, which I'd never been able to get to grips with. It's set out, with some mathematics, in Book III.

From the late 1990s, the standard model of cosmology, the lambda cold dark matter (λCDM) model, has been getting things right. It gives quite a range of numbers as it should, and a lot of physicists are sure it's correct. But since it appeared, measuring instruments have got more accurate, new surveys have tested it, and recently the cracks have started to show.

Since 2016, measurements of the expansion rate have shown that even with some fix-up type adjustments (involving unknowns), that we choose to put in, the present picture *still* doesn't fit together. There has been a discrepancy between two different ways of estimating the expansion rate, known as the 'Hubble tension', and it seems that yet another adjustment is going to be needed.

Since then, the discrepancy has worsened, and both groups are more certain of their results. The local value for the Hubble constant H_0, which gives the expansion rate, is measured as 73.5. CMB measurements, using the λCDM model, give 67.8.

And the method by which the apparently increasing expansion rate was first discovered, which since '98 has had many people certain that dark energy was pushing the galaxies apart, was itself thrown into doubt in 2015, when some astronomers discovered that the 'standard candles' being used, type 1a supernovae, were not so standard after all. The further away ones are like a different population from the nearby ones. They found this by looking in the ultraviolet range, and having found the pattern there, they then found it in visible light as well. But the whole idea was that these flashes of light stay intrinsically the same at any distance, and all our estimates are made from that assumption.

It's too early to tell if this will remove the anomaly that people put down to dark energy, but it's being said that it will reduce it - by an as yet unknown amount. But even if it leaves some of it, there may eventually turn out to be several populations of supernovae out there. Something similar happened in the 1950s, when cepheids, the standard candles for nearby distances, split into two populations, and estimates were found to be way out. Andromeda was several times further away than had been thought.

But finding that our measurements are out in this sort of way could actually resolve the Hubble tension. That part of the problem might go, but there are other parts of it, and the success of the λCDM model can be deceptive. It was picked out of a huge number of possible models, and fine tuned to fit the data in places. So yes, it works well - everywhere except where it doesn't. The problems have grown in recent years, and according to a New Scientist article from November 2020, people are seriously considering abandoning the λCDM model, and looking at gravity theories to replace GR.

But I don't believe one should be too critical when people are struggling with something as difficult as cosmology. And I don't do what some critics of the mainstream do, and just say 'therefore the entire present picture is wrong'. People often think a picture must either be totally right, or totally wrong. But it might be neither. It might be incomplete, still missing some pieces, wrong in places, but getting there slowly. That's what I think it is, and I wish them luck on the road. It's a hard road, but they've made real strides recently, and new technology that will help is on its way.

But it has to be said, we may be further from a complete understanding than many realise. Nowadays, when the picture we have doesn't add up, people very often think it needs just a tweak here, or a tweak there. I think instead it might need fifty or a hundred years, perhaps even centuries, before we fully understand the issues. In physics we understand a lot, having been able to check our ideas more efficiently as we go. But (setting my work to one side, and whatever effect it might have, if any), to me our physics is in some ways a few centuries ahead of our cosmology - that is, in terms of how complete our understanding is.

But the tendency is to work from the idea that the picture is nearly finished, needing only final tweaks. That assumption is not necessarily dependable or realistic. Several unknown effects may combine to make the problems we're up against - perhaps one or more unknown redshift effects - and to identify and disentangle them might take time.

Looking on the bright side, we're exploring new wavelengths, and there may be a series of breakthroughs. It's never easy to say how far away solutions to current problems might be. But finding them becomes more likely if people are open to the possibility of new, unknown physics, rather than constantly trying to patch up existing physics, and assuming it's all we need. It may be that our cosmology really is a few adjustments away from real understanding - but we don't know that. And I'd say the more we're open to genuinely new ideas, the higher the chances of getting there quickly will be.

89. The mass discrepancy

The mass discrepancy is 'the' problem in physics and cosmology. It's such an important problem that I'd talk about it even if I didn't have a solution, and in early drafts of the book I didn't - but this chapter was here nonetheless, in unresolved form, steadfastly exploring the possibilities.

I call it the mass discrepancy because we need to start with no assumptions. If we define a puzzle wrongly, we may already be walking down a cul-de-sac. So we can't call it 'the search for dark matter', and even 'the missing mass' is a name that might limit our perspective too early.

I studied the puzzle for many years without a solution. The one I have now is recent, and has not yet stood the test of time, but a lot of things clicked into place when I found it, and I absolutely think it's right. In particular, it explains two sets of clues that seems to contradict each other, as any solution for the mass discrepancy needs to do.

The puzzle is this. When we estimate the amount of mass in the universe via visible light, it's a far smaller number than the one we get when we estimate it via the gravity we can see at work, using our present theories.

So there are two possibilities. Either:

A. *There's an enormous amount of mass out there that doesn't give off any light (that we detect),*

or

B. *Gravity, for which we've only tested our theories nearby in our own solar system, works differently over large distances, or in different domains.*

The interesting thing about this puzzle is that the two possibilities *both* have strong evidence that appears to support them. So any solution will have to explain two seemingly contradictory sets of clues.

The majority of physicists think dark matter exists. But the idea that gravity needs to be modified has more support than before these days, in particular in a theory called MOND (modified Newtonian dynamics). MOND doesn't come with any explanation, but since 1983 Milgrom's theory has been slowly gaining support. It's still the underdog, but it's no longer a weird peripheral theory: good physicists take it seriously nowadays, and it has been said that as many as 20% of physicists are behind it. In fact it's not that simple, as a lot of people are unsure.

The idea is that with very small accelerations - the kind you don't find in the solar system, but which you do find in the outer regions of galaxies - gravity goes by slightly different equations. The adjustment is simple, and looks at first glance like a minor one, but it makes a big difference.

Stacy McGaugh's work applying MOND to large amounts of data has been very helpful to those who study these questions. He has shown MOND to work in quite a few different areas, with a wide range of ability to remove the mass discrepancy - without explaining it. MOND is not very adjustable, and can't be altered much to fit the data, which is an advantage. Dark matter is a far more adjustable theory. And loosely speaking, the more a theory has been adjusted and fixed up, the less likely it is to be right.

90. Evidence that points both ways

But the reality is, there's a lot of ambiguity in what we've been finding, using our amazingly good new tools for astronomy. Looking at the overview, it's far from clear. There's evidence for MOND, against MOND, for dark matter, and against dark matter. There's also evidence for modified gravity that is *not* MOND, and for dark matter that is not standard dark matter.

It's worth remembering that flat rotation curves, in which everything travels around at the same speed (at all distances from the centre), have now been found in clusters, and well as in individual galaxies. In clusters, the galaxies also orbit, and however they do it, they do exactly the same trick.

MOND and dark matter both have trouble in clusters. With MOND it's one of the problem places, by a factor of two. And standard dark matter is out by a factor of ten, on how much 'strong lensing' of light from small lenses within a cluster there is, from a 2020 study. The real universe seems to bend light off its path far more than with the detailed λCDM picture.

In fact, λCDM has been in major trouble recently. The results of a survey, mentioned earlier, shows that the distribution of dark matter is considerably too smooth: the universe is 'less clumpy than our best theories suggest', as someone put it. This is the latest of a series of blows to the central theory, λCDM, which includes general relativity. It's a particularly serious one - some have said it may be the last straw.

Some other kind of dark matter is now a better bet. There's baryonic dark matter: invisible normal matter. But that goes against theories that are seen as unshakable, and many are reluctant to look in that direction. But in 1999, two Dutch astronomers detected HII (H2), molecular hydrogen, in an edge-on galaxy, in the right amount to explain the galaxy's rotation. HII is invisible, and hard to detect. In other words, it's dark matter. Some think there's a lot of HII out there. The idea that HII is dark matter doesn't fit with certain parts of standard big bang theory, so the discovery was ignored.

But baryonic dark matter (HII, black holes, or whatever) is another avenue. It would mean some of our basic ideas are wrong - but then it's hard to find a scenario that *doesn't* imply that in one way or another.

People sometimes start from the assumption that dark matter exists, and then if there's a problem, they adjust it. Recently dark matter was suddenly given the property of being able to cool hydrogen - although it's not meant to interact with other matter - when a measurement suggested hydrogen in the early universe was too cold. The measurement was later contradicted by another one, so dark matter changed back again. Adjustments like these are often fine, but at times dark matter is like a fix-up theory.

Observations of the massive collisions between galaxy clusters have seemed to support dark matter strongly. In 2006, studying one cluster collision using the Chandra orbiting X-ray telescope, it looked as if the dark matter gets displaced from the other matter, and its gravity reveals its presence, via the lensing of light coming from behind.

Some called it 'proof of dark matter'. But since then we've been looking at other cluster collisions visible out there, and from 2007 onwards there was a major mystery. It suggested that our assumptions about this may be wrong somewhere. Another cluster collision also seemed to have dark matter in it - perhaps - but if so, behaving very differently. It seemed that the required properties of the stuff are different in the two scenarios. So at times, dark matter looks like some kind of universal 'polyfiller' used to fill the gaps in our knowledge, but needing nebulous, varying properties in order to fit them all. On the other hand, dark matter might be right - perhaps we just don't know how to interpret cluster collision data yet.

In 2012 two separate studies on the second cluster collision gave different results. One paper's title included 'the mystery deepens'. Later work largely reconciled the two studies, but it means the mystery remains. Astronomers have been truly stumped on this for several years now, and the problem is still there. All this makes caution advisable - we may be drawing premature conclusions on these questions.

The first of these two sets of cluster observations seemed to rule out MOND. Although some said a version of MOND called TeVeS can explain the data, few believed it, and the arguments for dark matter were very strong. But the problems with the second set of observations then tended to weaken that approach, and later studies of the first collision showed we had the speeds wrong, and drew some false conclusions that way. All in all, it seems there's something going on that we don't yet understand. Studies of dwarf galaxies around Andromeda seemed to rule out dark matter. But later, the way dwarf galaxies orbit together in flat disks turned out to go against not only that, but pretty much all other gravity theory we have.

In 2016 Stacy McGaugh and his team reported the result of many years of work. They had studied 150 spiral galaxies including a wide range of different ones. And they'd made an enormous discovery. They had found a very tight correlation between the flat rotation curves, and the radial acceleration that was predicted by the observed distribution of the ordinary, visible matter. As Stacy McGaugh put it: "*If you measure the distribution of star light, you know the rotation curve, and vice versa*".

The discovery took the form of a universal curve, which is generally found in the outer regions of many different kinds of galaxies. There's an often linked-to graph, with Stacy's 'radial acceleration relation'. The team didn't suggest a cause, but the point is, dark matter wasn't involved. Whatever causes those unexplained flat rotation speed curves, it seems to come from the ordinary visible matter, because there's an unexplained mathematical link, involving the acceleration, which doesn't need dark matter at all. This suggests MOND, but also any physics that doesn't use dark matter. Stacy's discovery dragged me back into studying the mass discrepancy closely, looking for any light that PSG could shed on it.

There's a very good potted version of some key arguments for MOND, to be found on some pages by Stacy McGaugh. It's in the reference section, for this chapter. It sets out, in shorthand form, five 'laws of galactic rotation', all of which are correlations with visible matter. But if I was asked 'do you think MOND is right?', I wouldn't be able to answer the question. It's just a bit of mathematics: what's needed is a conceptual explanation. Once you have the concepts, you then start looking for the mathematics, and part of what you find, when you find it, might look like MOND.

Stacy McGaugh has done something truly important. He showed that there's something very general going on. Within individual galaxies, there are close mathematical links between the visible matter and the extra gravity, leading to a neat description that simply doesn't need dark matter at all. For a while I saw this as clinching evidence that instead of dark matter, the cause is a key difference to gravity, in some domain Sir Isaac never considered.

But later on I looked at the evidence for dark matter from weak gravitational lensing, and it was like finding clinching evidence the other way. So it started to become clear that there are two sets of clues, and they seem to contradict each other. They're both directly related to the question, and neither can be dismissed. So the answer must somehow explain both. Needing to explain both is a restriction. But if you genuinely want to solve the puzzle *whatever the answer* (as some do), a restriction helpfully narrows things down.

91. A possible answer

A lot of people think the solution will be dark matter or modified gravity. It looks like one or the other, but it might be neither, including a bit of both. It's possible that what we call 'dark matter' exists, but with properties that make it far less like ordinary matter than we imagine it to be. It might even be stuff that we wouldn't call matter at all.

One day I realised that in the first cluster collision, the cloud of 'dark matter' that seemed to get separated from its cluster, and was seen via gravitational lensing, could perhaps be a cloud of secondary vibrations. According to PSG the secondaries fill the clusters, they create gravity, and they would indeed lens light from behind. They would come with the required density gradient to do that.

At the time I didn't think much about it, apart from thinking that I didn't have to worry about the measurement! With hindsight it was stupid, and I should have focussed on it longer, but this was before I'd looked far into the puzzle. I was still looking for a PSG modification to gravity that would solve it. It later became clear that it's a place where I missed something easy to see, of which there have been many. This was a particularly obvious one, but as discussed in Book I, expectation can affect what you see. And what I was expecting was modified gravity, due to apparently clinching evidence.

But later on, after getting nowhere with that approach, I went back to the place where an idea had briefly jumped in and out again. And thinking about it, a growing number of things started to click into place. The link that Stacy found, between the visible mass in a galaxy and the extra gravity, looks like it must mean modified gravity. But perhaps it tells us something else instead. It might mean that dark matter exists, but is somehow so closely connected to the visible matter that they follow tight mathematical patterns, dancing to the same tune. So the real value of MOND may be as a set of clues.

The connection is so very close that dark matter isn't even needed in Stacy's description - it's replaced by a bit of mathematics. MOND does the same, and Stacy uses a number that comes from MOND in his calculations. And yet in other areas the evidence for dark matter is very strong. So if dark matter exists, it would almost be like a part of the visible matter.

And that unexplained connection would also fit well with some other things that are known about dark matter. In maps that have now been made of the wider universe, at the scale of the giant threads of galaxies and clusters, the dark matter follows the visible matter, and clearly has a close relationship with it, but one that's not understood at all.

In my picture of gravity, a mass emits a refractive medium, which pours out of it in all directions, creating its gravity field. The medium it emits is made of very small vibrations in an underlying medium, which is the dimensions. In a gravity field like the Earth's, the secondary vibrations move outwards, and each part of the field keeps being replaced, so the field remains stable. This picture of a nice, simple spherical mass reproduces Newton's and Einstein's gravity. The waves travel away, dissipating as they go. Beyond that, no-one worries about where they go, or what happens to them. I never did, and no-one else did either.

But later I started reading up on what has been done with weak gravitational lensing, creating maps of the dark matter at the largest scales. If the cause of the discrepancy is a difference to the gravity equations, all we've mapped is what the extra gravity does. But the results don't look like that *at all*. They look like a map of dark matter. You get long filaments and threads of dark matter, with galaxies along them, and clusters at the intersections. A study in 2012 of one particularly large filament suggested it contains about 81% dark matter, 10% visible mass like galaxies, and 9% gas. An article said:

Mark Bautz, an astrophysicist at the Massachusetts Institute of Technology in Cambridge, notes that astrophysicists do not know precisely how visible matter follows the paths laid out by dark matter. "What's exciting is that in this unusual system we can map both dark matter and visible matter together and try to figure out how they connect and evolve along the filament," he says.

There's nothing wrong with not understanding something, it's exciting, and as in this example, physicists are often open about it when they don't. The patterns are at the largest scale, so they're a major part of what's happening in the universe. Although there's no understanding of them, the dark matter is definitely thought to come first, creating a web-like structure early on in the universe's history. The galaxy clusters are thought to come second in the cause and effect sequence: they're believed to form somehow at the places where there's more dark matter - at the intersections between the filaments. Somehow, matter then manages to 'evolve along the filaments'.

But suppose that approach has cause and effect the wrong way round. What if the galaxy clusters come first, and then (to put it in informal, non-technical language) they ooze goo. The waves that make up the field of each individual mass, where there are billions of masses gathered together, might create an excess, and build up into a huge cloud.

That's the price of having an emitted refractive medium in your picture (but I don't feel I need to apologise). If this is what happens, the dark matter in the filaments seen on our map may be enormous amounts of refractive medium, pouring slowly out of each cluster - I say 'slowly' because the picture is a large one, but they travel at c - and drifting slowly along enormous bridges of matter towards other clusters.

They would only have gravity if the cloud has a density gradient, but a cloud such as that would tend to have one. It's not just that each individual mass creates a gradient, as its cloud dissipates at the edges. It's more about the way matter is at larger scales: individual galaxies, clusters of galaxies. They have a gradient, so you get a cloud with one. And it's *also* always thinning at the edges, and always being replaced by more cloud arriving behind it. And always moving, because the secondaries are very like light - they never stop. So they flow along filaments of matter and flowing gas.

In the 'missing mass' problem we assume the gravity we measure is directly associated with a mass, and that we need to identify this mass. In PSG it isn't, so perhaps we'll find that the missing mass is no longer missing.

And the flat rotation curves in galaxies, which are so challenging to explain, might arise because the RM cloud moves outwards from the centre of the galaxy, distributing itself so that the density gradient (which is what causes gravity) creates the pattern. The further out, the more extra gravity, as if the total mass increases linearly. One of the first questions is, can this reproduce Stacy's universal curve? That's what any solution needs to do.

92. A medium that dissipates?

As I said, the λCDM model got a lot of things right, over 20 years. But better measuring instruments, and results from long duration surveys just in, have started to show major cracks.

And recently it has been getting harder to move the standard view around to make it fit, which people did in the past. There are now two discrepancies that are separate, but indirectly connected, because if changes are made to reduce one of them, they tend to make the other worse. One is the Hubble constant discrepancy, which I've mentioned.

There's a 2019 study that points this out, and suggests an unexpected way to ease *both* problems, with what's called a 'decaying dark matter model'. The idea is that both discrepancies can be reduced by assuming that dark matter isn't long lasting in the way that ordinary matter is. This is interesting, as in PSG the secondary vibrations dissipate as they travel.

I've not looked far into their model, but there may be similarities. According to PSG, the substance they're looking for isn't matter at all, and has no mass - it's more like light. But the physicists who found the decaying DM model say that the mass doesn't affect things in their picture: *"As our analysis here is phenomenological, our constraints are independent of the mass of the dark matter particle undergoing decay."*

Back with galaxies, what happens if not much excess medium has built up? It has been found that dwarf galaxies lack dark matter. They're small, and PSG says they haven't emitted much of the stuff. In 2022 it was found they get pulled apart, and warped, too easily when put in a cluster, as there's not enough holding them together. This doesn't fit the λCDM model.

There's one more clue, which suggests, if nothing else, that dark matter does some strange things, and is connected with gravity itself. There have recently been observations that show galaxies at large distances apart moving as if invisible connections linked them. The distances are too large for gravity as we know it to be the cause. It's as if filaments rotate, and carry galaxies with them. 'Dark matter' as in PSG - emitted from clusters - might rotate, as that's what clusters themselves do. A 2019 article called 'There's growing evidence that the universe is connected by giant structures' looked into it. Since then the fact that filaments rotate has been confirmed.

93. A new way forward

If this general solution from PSG is right, in the longer term it would contain the seeds of a switch away from general relativity. If so, physics would need PSG to move forward, because of its unique conceptual elements. If it *is* the secondary vibrations doing it, and if that turns out to fit the observations and data well, then theories other than PSG will not solve it in any complete way, and the mass discrepancy would start to look like evidence for PSG.

It would fit with the fact that we can't find any dark matter, after decades of trying. And yet we find enormously strong evidence that it exists. According to PSG, what we call dark matter isn't matter: the stuff we've been looking for is made of waves at the very smallest scale. They can't be found with any of the indirect methods we use for detecting dark matter.

Stacy McGaugh has shown some evidence for an all baryonic universe, with no matter that's any different from ordinary matter. If dark matter exists, but isn't actually matter, you can have *both* what we call 'dark matter', and an all baryonic universe, in a cake-and-eat-it kind of way.

So there you have some clues, and it's clear that the emitted medium in PSG is a good candidate for dark matter. It ticks some the main boxes: you can't see it, and it's only detectable via its gravity. It generally doesn't interact with ordinary matter in any other way. It has a very close relationship with visible matter, which we know about from mathematical links between the extra gravity and the visible matter in galaxies. It has this relationship because the visible matter emits the medium in a consistent way.

But what would it *do* that's different from the standard idea of dark matter? Well, that's what I've been looking at, and it's early days - I have a long list of ideas. It's too early to say very much, but it looks like the refractive medium attracts matter via a kind of gravity, but not itself.

Doesn't the medium have its own basic gravity? No - the secondaries are like light. They have energy but no mass, just as light does, in both GR and PSG. (In GR, technically, light can be said to have gravity, but not in PSG.) So the RM, like light in PSG, has no basic gravity. But unlike light, a density gradient can give it another kind of gravity, which affects matter. And like any group of waves moving together, although they don't have self-gravity, they have a different kind of cohesion, involving short-range forces.

So the RM might be modelled as non-self-gravitating 'dark matter', located and distributed consistently with it having been emitted by the visible mass. That's a possible solution for the dark matter puzzle, what about the 'dark energy' one? A solution for that, and for the 'vacuum catastrophe', are both in Book III, which has the background picture for the whole theory.

And synchronised satellite galaxies? We find there are flat disks of corotating miniature galaxies around larger galaxies 'throughout the local universe', as one paper put it, although they're expected to do it in only 0.2% of these situations. They somehow rotate together. This was seen around our galaxy in the 1970s, then Andromeda, and recently many others. So far, we don't even begin to begin to explain that. It goes against GR, the standard λCDM model, and more or less everything else we have.

So a *very* out-of-the-box idea might be needed. With this starting idea, it may one day be possible to explain it: dark matter is itself the gravitational fields of those galaxies. They emit a medium, so perhaps the more satellites there are in that plane, the more a disk of it builds up, bringing in more galaxies, in a feedback loop. It's still early days, but this new picture creates some much needed legroom for new ideas.

An in-depth part of the book follows, with more detail in places.

Part 17. More about the flyby anomaly

94. The best clue of all?

So now we return to the flyby anomaly, and some more clues. Some of these points were extracted by Luis Acedo from the NASA data about Earth flybys, which was made available online. He put a lot of graphs in his papers, which revealed things the Anderson team didn't mention.

Like the NASA team, Acedo and his colleagues saw the anomaly as to do with the planet's rotation. Bel saw it as also related to a retardation of the effect of gravity, and he worked with Acedo on that line of thinking. In GR gravity officially travels at lightspeed rather than being instantaneous, but it never normally makes a difference, and there's no noticeable delay. Bel thinks that for some reason in this case it does make a difference.

But this view can only explain some of the clues, not all of them. So I should set out a few. There's one point (simply referred to in my notes as 'the weird clue'), which is almost unavoidably important - it could be the best clue we have. Acedo says that Anderson also mentions it.

When the probes speed up as they pass the Earth, one might wonder if the Earth's rotation is pulling them around slightly, as they fly nearby it. Perhaps the probe somehow attaches slightly to the Earth when it gets near enough, and the planet's rotation pulls the probe around with it a little.

But that's not the case. Most of the anomalies are speed increases, but a few have been decreases in speed. And Acedo and Bel point out a pattern, in a paper *'On a correlation between azimuthal velocities and the flyby anomaly sign'*. Azimuthal means, loosely speaking, in a direction parallel with the equator. 'Sign' of course means simply plus or minus: whether the anomaly takes the form of an increase or a decrease in speed.

What they found is that it's not what might be the expected pattern, it's the other way round. When a probe flies near the Earth, and travels around it in the same direction as its rotation, there's a *decrease*, not an increase, to its speed. So the Earth isn't pulling it in a straightforward way. But flybys that go against the Earth's rotation show the opposite: a speed increase. Sometimes no anomaly is measured, but when there is one, that's the pattern.

Now this is probably a major clue. It shows a clear connection between the anomaly and the planet's rotation. And although it's not the expected kind of connection, in which the Earth drags the probe around somehow, that might actually help in an odd sort of way. It rules out a lot of possible causes, all of which would go the other way. So although it still leaves a baffling mystery, it helps to clear the fog, throwing out a lot of possibilities.

Could the pull of gravity be at a slightly different angle - in a slightly different direction than the radial? It might be that when a mass rotates, the pull of gravity is in a direction angled slightly against the planet's rotation, that is, pulling backwards. This possibility could fit with the clue that Acedo and Bel pointed out.

If so, what, according to PSG, would be happening? Well, it may be that the secondary vibrations are being thrown 'forwards' slightly, in the direction of the rotation, as they are emitted. That would mean the secondaries *that the probe encounters* would not have come from radially below it. Instead they'd have come from behind it if it's moving with the Earth's rotation, and from ahead of it if it's moving against the Earth's rotation. So this potentially alters the direction of the pull, to slightly against the direction of rotation. It would do this in one way or another - could the angled emission affect the direction in which the medium changes in density fastest, because the field is 'packed' in a different way, and at an angle? Not necessarily.

But there's another possibility, which I won't go very far into here. It's more speculative in the context of PSG. I've assumed for years that the direction of the pull of gravity is always simply the direction where the refractive medium density is changing the fastest. If so, that might mean towards the centre of the mass, even if the RM is emitted at an angle.

But speculatively, perhaps the emission of the medium at an angle (thrown forwards) creates a reference frame, because the secondaries like everything else travel, locally, along the dimensional cylinders. If so, perhaps close to a rotating mass, the pull of gravity happens in a direction that is along the local alignment of the dimensions, as implied by that reference frame.

To me this is an interesting idea, but adding in that sort of extra complication may not be needed to solve this (and I can see William of Occam frowning at me). As I said, I have no exact solution, so I'm just throwing out ideas - I hope they're of interest. And although we know that the probe itself doesn't get dragged around by the planet, perhaps the medium does.

I can calculate the emission angle, and have experimented with it, it's simple to do. The formula for that depends on a few things, but mainly the planet's surface rotation speed at the emission point. For that you need the latitude at closest approach, which can be got hold of for some flybys.

The trouble is, the latitude of closest approach varies, so the emission angle varies a lot along the flightpath. But the NASA formula for the anomaly some years ago showed that the difference between the incoming and outgoing angles to the equator affects things, and it's not clear if this can be related to that formula. If I start getting numbers out that fit, I'll send the formula for the emission angle to the excellent physicist who nowadays leads the search for a solution, Luis Acedo. You never know, perhaps when he - or someone else reading this - looks at it, a lightbulb will come on.

95. Closest approach

The other clues I have for you are about what happens at the point of closest approach, or perigee. It seems to have something special about it. For a long time the probe moves closer and closer to the planet. After closest approach, it slowly moves further from the planet again, doing what is more or less the reverse. But at the point of perigee, when this process turns around, we find that some strange, unexplained things happen.

In Acedo's 2017 paper *'Anomalous accelerations of spacecraft flybys of the Earth',* he shows some graphs of the anomaly, comparing the NASA data with his own model. The graphs deal with quantities such as the probe's radial position, its radial acceleration, and so on.

The period of time that the graphs cover tends to be from 15 or 30 minutes before closest approach, to 15 or 30 minutes after it. So it focusses on that particular moment, and for good reason. (Some graphs do a few hours each side, as in a later paper on the Jupiter flybys.) But the point is this: whatever the graph of the anomaly is doing before and afterwards, it always goes to zero at closest approach.

The line on the graph might be on a slow downward diagonal path from left to right, and it might continue the descent afterwards. But right on closest approach, the graph briefly levels off, and you see a little flat ledge, meaning zero: no anomaly there.

With the radial part of the acceleration, as Acedo points out, for Earth flybys the *sign* of the anomaly changes at closest approach. This is odd, and hard to explain. It flips from an outward to an inward acceleration, or vice versa, and carries on that way. And he shows graphs for five of the flybys of the Earth -

in some it flips one way, in others, the other way. But the anomaly is *always* at zero, or very near it, when the probe comes closest.

So what's so special about closest approach - why does the anomaly go to zero there? At that point where there's no anomaly, a probe behaves exactly as it's expected to in Newton's gravity theory, or Einstein's.

Why? Well, perhaps because it's moving horizontally (tangentially) in relation to the planet, rather than descending or ascending, which is what it's doing just before and afterwards. And the probe's path in relation to the planet is briefly more symmetrical than it was before or afterwards.

There's a graph for the NEAR probe's radial distance from the centre of the Earth, which clearly shows what looks exactly like retardation. There's a time delay between two almost identical curves, for the expected and measured behaviour.

Acedo amplifies the delay by a factor of 1000 on the graph, so it can be seen. It's clearly visible both before and after closest approach, and in both cases *the delay increases with distance from the mass*, as if gravity was taking time to get to the probe. If you calculate it from that graph (Figure 1 in the paper), you get *c*, a lightspeed delay, for any point on it. For some reason the probe's position, although it changes over time, is always late, compared with the expected position. And it's always late by the time it takes for light to travel to it from the planet.

So we have a lightspeed time delay between the expected and the measured behaviour. Figure 1 of Acedo's 2017 paper shows it beyond doubt. And that, as I said in an earlier chapter, means that something associated with gravity, but which probably isn't gravity (or the solar system would collapse), travels at lightspeed, and affects the probe.

The graphs for Juno's orbit of Jupiter are a little different, as it's an eccentric elliptical orbit, not a hyperbolic one. The paper, *'A possible flyby anomaly for Juno at Jupiter'*, has more clues and graphs. In one, the radial acceleration is wildly high just before and just after perigee. But the wild zigzagging around near to perigee was almost identical on two different orbits Juno made. And right on closest approach, in both flybys, the line dives straight down to zero, then straight up again. No anomaly at closest approach.

I don't have explanations for all of these clues, but it may well be that we're seeing the behaviour of the emitted refractive medium from PSG. As I hope you'll agree, there are strong hints that seem to suggest it. There are many clues, and they suggest complex, sophisticated behaviour, perhaps like that of a swirled emitted medium, with outward motion, *plus* rotational motion. And the rotational motion varies with latitude - the planet turns faster as you move from pole to equator. This would explain why latitude comes into so much of what we find on the anomaly.

It's clear that from the probe's point of view, the configuration of the field is changing over time. And in PSG the medium does indeed have a lightspeed time delay. Perhaps that makes a difference that sticks out in asymmetrical situations. After all, the medium is what tells the probe how gravity should affect it.

The clues I've just given you may or may not be interesting. But the anomaly is on the frontline in the battle to understand gravity, and the latest part of a long detective story. We've been chasing gravity for 300 years or more, and although it seems that we're not there yet, we're going to get to the answers in the end.

There's not much doubt that the anomaly is about gravity. Some are trying to solve the flyby from within general relativity, as part of a hidden retardation effect that GR gravity contains. Others are trying to solve it from outside GR, and PSG is one of many approaches. But either way, these points and clues are part of the recent developments in the ongoing story of humankind and gravity, and it's not over.

96. The increase to the Earth-Sun distance

There's an increase to the Earth-Sun distance, happening steadily over time, and it's another mystery. It's a small puzzle so far, so it's tacked on at the end of this section. It's often referred to as the secular increase to the AU, but that risks being hair-splittingly wrong, given how the AU has been defined since 2012. The puzzle was discovered in 2004, and the people who found it have ruled out quite a list of possible causes.

There's also an increase to the Earth-Moon distance. We've bounced lasers off the moon for decades, as the astronauts left reflectors behind there in the 1970s. So we measure the Earth-Moon distance with great accuracy, and it's slowly increasing, or the eccentricity of the Moon's orbit is. Now with the Moon, there are a lot of subtle effects that might be contributing. But with the Earth and the Sun, there are fewer possible explanations. But there still are some, and it doesn't have to be new physics.

On the other hand, it might be. In the PSG picture, it might be the secondary vibrations. Although they cause gravity's pull on matter, paradoxically they might also have an ability to push matter away. It's a far smaller effect - the secondaries are very small, but they pour out of the Sun in such enormous numbers that they might gently push the inner planets outwards over time. I don't know if they cause this anomaly, but if so, further work might lead to mathematical evidence for PSG.

But there's another possibility: it seems very possible that PSG orbits change slightly over long periods of time. The vis viva equation was arrived at via the conservation of orbital energy, and helical refraction leads to numbers that are different after nine decimal places. That might mean the new, real orbits are less stable than the old, false ones, which could explain some things that we've observed recently. But to see what PSG orbits do over long periods of time, further work is needed - and a sophisticated computer programme. Also, other orbiting systems will be observed closely soon, such as pulsars, where the differences between GR and PSG may be visible.

Returning to the puzzle of the increase to the Earth-Sun distance, anyone looking into it needs to take something else into account. Since the mid '90s exoplanets started being discovered, orbiting nearby stars. There are several thousand now. PSG might explain why they're on very eccentric orbits more often than expected. But some are Jupiter-sized planets on very close orbits around the star, doing one orbit in just a few days. We don't know how they migrated or bounced inwards, but they couldn't have formed there.

It's not surprising that we found those planets early on, because our system for finding exoplanets selects them, and particularly now, when in its infancy. They're the ones that are easy to find. But it's a mystery how they got there, and the point is, if they move steadily inwards for some reason, as is widely believed, we'd then need to explain why the Earth is instead moving steadily outwards.

So there may be two effects at work, each partially masking the other. Which one is larger might depend on the situation. And that's one too many effects for it to look like a promising avenue, particularly when there are so many other worthwhile avenues to explore.

Part 18. More mathematics: orbit corrections

97. Post-Newtonian adjustments

Just as GR puts minor adjustments on Newton's gravity equations, PSG does something similar. But instead of being applied to Newton's equations, the corrections go onto PSG's own equations, which are already slightly different from the standard ones in places. So this takes things two steps away.

With speeds for orbits, what happened was that the corrections were found before the actual equations. From very early on, it seemed that whatever the orbital speed, the refractive medium would have an extra slowing effect. The orbit corrections sometimes affected things just as in GR, but it depended on the direction in which an object was moving.

Picturing a gravity field, for simplicity one can imagine a flat two dimensional diagram. There are two basic directions: the radial direction is straight lines outwards from the centre, and the orbital or angular direction makes circles at different distances from the centre. It's often best to call it the non-radial direction. Across a small distance it makes a horizontal line, but over a larger distance it's an arc or a circle.

All orbital motion can usually be described using these two directions. And in GR, where curvature affects things is in the radial direction. That's where the adjustments onto Newton's equations come in - when light or matter moves between different heights, towards the mass or away from it. If you imagine the field in GR as a dimple in space, the radial part of an object's motion, if it's on an elliptical orbit, is about when it climbs up or down the 'slope', on its way inwards or outwards from the mass.

And when it climbs either way in the radial direction, it seems to slow down very slightly from the Newtonian speed. That effect can be taken in different ways in Einstein's theory: one way to take it is as a stretching of measuring rods in the radial direction. But in the picture of GR I've used in this book, it's an apparent slowing we see in our three-dimensional view on the world - in a fourth direction the object travels around the invisible curve of the crater, where there's no slowing effect, but further to travel.

So the radial direction needs these adjustments in GR, to add the curvature. But perhaps surprisingly the non-radial direction is unaffected: circular orbits stay the same. An object on a circular orbit doesn't move radially, it stays at one height some way up the sloping wall of the crater. It then travels around like a motorbike stunt rider - the plane of the circle is slightly offset from the flat in a fourth direction, but that's not observable. The curvature of space makes more or less no difference.

So a GR circular orbit is like a Newton one. The post-Newtonian adjustments, when they arrived, left circular orbits alone. Apart from a few things like time rate differences, they're just as they have been for three hundred years.

98. A difference?

So in GR, loosely speaking, radial motion needs the slowing factor, but orbital motion doesn't. (This is an approximation - in GR a 'diagonal' path across the field is more complicated.)

but PSG is different. In PSG the equivalent slowing is a simple, literal slowing in three dimensions. It's exactly the same in all directions: at a given point in the field, the medium slows any object in the same way, whatever direction it's travelling in.

So a rare difference between the two theories appears. The theories are the same in the radial direction, different in the non-radial direction. In PSG you get a special extra slowing in the non-radial direction, which affects any orbit near to circular, and the non-radial component of all orbits.

This seemed exciting when I first found it, but the difference remains hidden in most situations - it disappears into the size of the planet's mass. The value of the central mass affects the orbital speed, but we usually don't know that number accurately, *except via orbits*. So say the adjustments of GR applied in the orbital direction as well as the radial one for some reason, how would we know? With a circular orbit, if you add that in, all it does is increase the size of the central mass very slightly, to compensate.

To put it another way, when observing a particular circular orbit, there's no way to tell if it's a PSG orbit around a slightly larger mass, or a standard orbit around a slightly smaller one. The difference is so small, it's not necessarily easy to check that via another route.

I found this out in around 2007, which was early on. It looked like a key place where the theories diverged, but PSG was incomplete, and lacking the orbital speed itself. At the time it seemed to me that the PSG adjustments would go on Newton's equations, like the GR adjustments. But I found out in 2011 that PSG has its own orbit equations, and that the adjustments go onto those.

An equation found in around 2007 (along with an earlier version of the free fall equation), gives the circular orbital speed with the slowing effect. It was reached by multiplying Newton's circular orbital speed by the expression for the slowing effect of the RM. Because that factor is the same at every point on a circular orbit, it's straightforward to include it. For the circular orbital speed, the derivation was simply:

$$v = \sqrt{GM/r} \ \times \ \sqrt{1 - (2GM/rc^2)}$$

so

$$v = \sqrt{(GM/r) - 2(GM/rc)^2} \tag{43}$$

This later turned out to be only part of what happens. In fact the corrections go onto PSG, rather than Newton. But this was about circular orbits, and it turned out that in that instance, the result is almost exactly the same. And it still showed an interesting point.

The difference isn't just about PSG. It's an aspect of the relationship between two basic principles: space curvature and refractive medium. Places where these two general interpretations differ are rare, and particularly important. But although it exists on paper, this difference is another of those elusive, hard to trace ones, and it usually stays hidden from view.

99. A cancellation

When the helical refraction equation arrived, the corrections were already in place, and understood. They involved multiplying all speeds, in all directions, by the expression that gives the time rate and the gravitational redshift. The RM slows all travel speeds, which was part of the original premiss.

Soon after that, perhaps a few weeks, the generalised orbital speed equation came out of the helical refraction equation. This was in 2011. The question then arose as to whether the corrections were still needed on top of it, and it seemed that they would be. But this wasn't obvious, because the corrections represent the effect of the RM itself, and that had been put in already. Using Snell's law, as it was used, certainly included the local RM speeds, so why put them in again?

In the use of Snell's law, the local RM transmission speed leads to the helical path angle. But there's also a *direct* slowing to matter in orbit caused by the RM, which is what the adjustments are about, and that's extra and separate. It's separate because it doesn't change the helical path angle, so it doesn't enter the Snell's-law-related side of things. It doesn't affect the helical path angle because it slows all three speeds on the helix by the same factor, so the sides of the triangles built into the helix stay in proportion, so the angle doesn't change.

So applying the adjustments on top of the orbital speed equation is required by the theory. But it also fits very well with measurements. It explains - for anyone who believes PSG is correct - why the GR adjustments have improved orbits as they have. By that I mean the relationship between the calculated orbits and the measured orbits. The adjustments can also explain things like the perihelion shift of Mercury - PSG closely mimics the GR adjustments, and any differences are often negligible or unmeasurable.

And what happens is you get an odd cancellation, which I'll show you. It even removes the difference to the mass I've mentioned. PSG, unlike GR, has the corrections in the non-radial direction. But by chance they tend to cancel the extra third term PSG adds into the orbital speed equation itself.

With elliptical orbits near to circular, which we deal with a lot, the extra term in the vis viva equivalent equation speeds the orbit up, but the RM slows the orbit down - by almost the same amount. The cancellation gets increasingly precise as orbits become more circular and less eccentric, and for an actual circular orbit it's more or less exact. So the end result is that the two ways in which PSG should stick out cancel each other on orbits near to circular, and any difference lands in the uncertainty margins of the central mass value. So PSG gets hidden, once again, in the data.

In setting the theory out, it often helps to add these adjustments separately, because the PSG equations have equivalents in standard theory, so for clarity they can be set out in the same sort of way. The result is similar to Newton's theory, with the GR adjustments on top of it.

But the complete orbital speed equation, RM slowing effect included, is very useful nonetheless. Because it works in all directions, orbiting objects in fact always travel at this speed, so it's important in PSG. I've set it out before, it's this:

$$v = \sqrt{GM\,(2/r - 1/a\, + [2GM/arc^2])\,(1 - [2GM/rc^2])} \qquad (42)$$

That applies in all cases, and gives the true speeds at which objects orbit. But in the special case of circular orbits, the cancellation can be seen clearly. First you look at how one gets to a circular speed equation. Two steps back from equation 42, adding in the corrections means that you have the PSG orbital speed equation, equation 26, multiplied by expression 4:

$$v = \sqrt{GM\,(2/r - 1/a + 2GM/arc^2)} \ \times \ \sqrt{1 - (2GM/rc^2)}$$

Because for circular orbits $a = r$, we branch off from that step in a different direction, and that goes to

$$v = \sqrt{(GM/r) + 2(GM/rc)^2} \ \times \ \sqrt{1 - (2GM/rc^2)}$$

and then

$$v = \sqrt{(GM/r)(1 + [2GM/rc^2])(1 - [2GM/rc^2])} \qquad (44)$$

I'll simplify this further in the next chapter, but you can see the cancellation I've been talking about. It's close similarity, in equation 44, between the second and third terms in round brackets (averaged, they simply disappear, but multiplied together as they are here, they make a number so close to 1, the two terms effectively cancel). This cancellation affects any elliptical orbit that's near to circular, such as inner solar system Sun orbits - Venus, Earth, Mars - and a lot of orbits around the Earth.

100. Some incorrect ideas

Many sensible physicists ignore attempts that are made by a small minority to generalise about gravity theories. One misconception that can accompany this is the idea that physics always reduces to mathematics - it doesn't.

Elements of theories often have to be applied specifically, which means that many theories are immune to attempts to generalise on theories en masse, without looking at them closely, made almost always by proponents of GR. A few people have tried to round theories up into groups, and herd them into categories.

When they try to do this, assumptions are made about what theories should contain, and it's thought that if they don't have the right ingredients in some pre-assumed way, they can't contain anything else to produce the necessary results, and therefore can't match observations.

This is a 'mathematics only' approach, which implies that there's no need to look closely at what a theory does to explain observations. The systems they use refer to Newton's theory and GR, but they don't refer to the conceptual side, even though there's a need to do so. For one thing, the conceptual side can imply different mathematics in certain situations, in a subtle way. That system is like a crude 'multiple choice' exam question, where the answer you need to put sometimes isn't on the list. And in general, attempts to bypass the conceptual side, as I hope I've shown, have been a major source of error in the past.

This approach even tries to make statements about theories that haven't yet been found, the physicists haven't yet been born, and although the parents may have met, and may like each other and fancy each other a bit, they've not even been on a date yet. It's an approach that effectively tries to make it difficult for a new set of ideas to challenge GR with an alternative. But the conceptual side may turn out to be far more important than this implies, and future generations may look back and find strange this 'mathematics only' approach that some physicists take at present.

Some have even tried to lay down fixed criteria for what acceptable theories of gravity should contain. But there's disagreement among them about what these criteria should be, and that includes serious disagreement between the two main people who try to do this. That speaks for itself.

One physicist decided that all gravity theories must be about fields, and tried to prescribe rules for what a gravity theory should contain from that. It's not only that he was trying to impose his own views onto other physicists' work. It includes a failure to accept the nature of physics, in which the equivalence we find everywhere means that new and unknown principles can suddenly come into play a few decades on, which no-one anticipated at all.

Incidentally, I've seen similar talk with quantum mechanics, where attempts have been made to define what new interpretations for QM should contain. This is often better, in that *explanations* are called for in some areas, but it's still unnecessarily restrictive. In my own interpretation for QM, many of the requirements that tend to be listed are there, but to find them I still had to go outside the basic channels of thought, and people sometimes forget that a new interpretation might remove some of the existing landscape, or alter parts of it, or redefine parts of it.

So trying to shoehorn future theories *into* the existing landscape using a set of rules doesn't work. You have to admit that the existing landscape might be flawed. Those who take this attitude are ultimately showing a resistance to new ideas, and refusing to admit that false ideas might be sitting there in our present landscape, that could invalidate the thinking that arises from it. Any refusal to admit this possibility is not good science, and it's an attitude that holds science back.

So if someone doesn't know how gravity works, or how quantum mechanics works, they're not in a position to say anything at all on what a description of how it works should contain. The reason is that if standard theory is wrong, *the assumptions surrounding it also become questionable,* so they can't be used to specify anything about that scenario.

This means that no-one can limit what might replace standard theory. Much of the authority that some have at present stems from the deeply-ingrained assumption that GR is right. Their authority rightly allows them to talk about GR, but they're not in a position to talk about anything unknown that might replace it - partly because being unknown, it's likely to contain things they don't know about.

And if GR did need replacing, those who had authority on it would no longer necessarily have any authority on gravity as a wider area. The authority that would take over would have to be the scientific method. Our task is to have an open mind, and the scientific method prescribes testing by experiment. Happily, this leaves some natural legroom for genuinely new ideas in science, which are science's lifeblood. The final arbiter about physical theory is, and always will be, experimental verification - not, for obvious enough reasons, a proponent of one of the theories to be tested, such as GR.

The only people who understand GR well are people who have studied it for decades. They do not have much to motivate them to allow it to be replaced, and they're less accountable, because no-one else can say much about it. And yet, naturally enough, the world only hears about certain things from them. So experiment, as always, is the only way to find out, and the next part of the book deals with experimental evidence.

Returning to gravity, PSG is loosely speaking a scalar field theory, but it has elements in strong gravity that make this aspect an approximation. Both the underlying equations and the post-Newtonian corrections are different from standard theory. This creates the cancellation I've shown, one result of which is that the circular orbital speed approximates Newton's theory very closely. Simplifying it further, with the corrections in, one gets:

$$v = \sqrt{(GM/r) - 4(GM/r)^3/c^4} \tag{45}$$

The extra subtracted term is what the cancellation leaves behind it. It's far smaller than all other small differences I've talked about. It's so small that it's undetectable nearby: it disappears into the mass. But PSG is self-consistent, and at the Schwartzchild radius equation 45 goes to $v = 0$, as does equation 42 for elliptical orbits. PSG says it's only relevant further out, and as we'll see in Chapter 110, equation 45 may already have been spotted at work.

But all else aside, for most purposes the circular orbital speed reduces to a familiar shape:

$$v \approx \sqrt{GM/r} \tag{46}$$

So as is often the case - from my point of view, and from most other points of view as well - Newton wasn't far off the mark.

101. Radar ranging data

Bouncing light signals off planets, and timing the echo's return, is like using sonar, and bouncing sound waves off the seabed to measure the depth of the ocean. This is one of many analogies, like the word 'spaceship', between the seas, the previous frontier, and space, the next - which some, of course, say is the final frontier.

In the inner solar system, orbit parameters affect each other, and they can't always be taken separately. Radar ranging is very accurate nowadays, but it's only used to get some of the numbers - others are reached from those, and the radar ranging data itself is interpreted assuming GR is correct.

Any differences due to PSG, which could in principle stick out here and there, are very small, and could get hidden. What the ephemeris software does is it assumes GR is correct, then makes a simultaneous fit, using all the measured numbers, the derived numbers, and the uncertainty margins. This method is capable of to some extent smoothing over any small differences, spreading them out across the uncertainly margins of many numbers.

This is not a problem if GR is right. The software paints the GR picture very well, but it doesn't search for anything else. Unlike humans, some robots can only read one theory. I'm not blaming the robots, or the humans (and there was never any kind of dispute or pitched battle between them). But there could be room for these minimal differences to slip through the net, without producing any glaring anomalies in orbits.

Anomalies would be more likely to appear in very eccentric elliptical orbits, such as the orbits of comets, and in hyperbolic orbits, such as the orbits of some space probes. In these there's no cancellation, but instead, where the path is near to radial, the GR adjustments land on top of the PSG ones. So that term is expected anyway, which removes it. But the PSG differences to the orbit speed itself, although sometimes lost into uncertainties such as the planet's mass, are then potentially exposed.

And in fact, anomalies have indeed appeared in those kind of 'long' orbits. The Pioneer anomaly looked like it might be the result of this, but trying to pinpoint the cause didn't lead anywhere. The differences between PSG and

GR were always four orders of magnitude (ten thousand times) smaller than the anomaly. They included seven or eight very small effects, and calculating what they do is a job for a robot. But one benefit of looking into the Pioneer anomaly was that I explored that whole area.

Comets arrive late sometimes, but that doesn't have to be a gravity anomaly. With comets it's hard to tell - no-one knows what they're doing. They go out into the Kuiper belt, where there are large chunks of ice floating around, and these might deflect the orbit path.

In general, Sun orbits are not the right place to find evidence for PSG, though they might be in the future. What about distances to the outer solar system? According to PSG they should be slightly different - by a few meters here and there. Quite a few spacecraft have been sent out there, and they find their way around remarkably well, but they would either way.

When a radio signal is bounced off an object, or sent to a probe that boosts and returns it, we work out its distance by timing the signal. Doing this in the outer solar system is different from the Mars time delay of light experiment. There we knew the travel time already, by sending signals to Mars when it wasn't directly on the far side of the Sun, to compare with.

But in the outer solar system we don't know the distances quite so well. We send signals out there, time the gap until they return, then use our theories to estimate how far they went before they turned around. The Sun's gravity field gets steadily weaker out there either way, but it still does what it does, whatever that is. And curved space and vibrating space give slightly different distances.

But each version of the solar system can be self-consistent in relation to the data. You can't bounce a signal off a gas covered planet, but a space probe can send one back. Estimates for their distances may work whichever theory one uses. It's possible to get a consistent picture, with the data interpreted differently - anomalies wouldn't necessarily show up that way.

One might think that orbits would give it away, if the distances were slightly wrong. But orbits are modelled partly via ranging data, and assumptions are used in interpreting it. Orbits also have built-in ambiguities - for instance, the eccentricity, which is important for distances, is a vague parameter. Timing an orbit doesn't tell you how eccentric it is, because for a given semi-major axis, the period is exactly the same for all elliptical orbits, however elongated they are. This happens to make distances harder to pin down. So there are numbers that get filled in via other GR-derived numbers, and in general the outer solar system is still, to some small extent, ambiguous.

What's more, an analysis has been done by the good Italian physicist Lorenzo Iorio of a related issue (he and I were both published in the same journal a while ago). The issue Iorio studied boils down to 'put GR in, get GR out' in the context of solar system experiments that have already been done, and other planned experiments. Although he identifies the problem only at a very high level of accuracy, he warns of a danger that the next generation of tests of GR by experiment in the solar system will become redundant within a certain range, as the results are beginning to be affected by GR-based assumptions that went in initially - those could include a whole range of numbers.

So there's room for other interpretations. A set of solar system parameters is like a landscape, or it creates one, and it takes the form it does because of the theory that shaped it. A different theory might lead to a subtly different landscape - but still coming out of the same data.

Part 19. Experimental evidence

102. Looking for a measurable difference

In the scientific method invented in the Renaissance, experiment became the final arbiter. Newton had more than a hand in that. It wasn't a lofty principle, he sat around at home building experiments, trying to find out for himself what was behind the physical processes around him.

He put two prisms together with one reversed, and found that light could be first broken up into a rainbow, then reassembled back into white light. This showed that white light contained all the different colours. People had done the first part of this for more than a thousand years, but no-one had thought of the simple, lateral second part of it. Putting the light back together again removed the ambiguity, allowing Newton to find out a lot. He also invented the reflecting telescope, and Galileo before him had been one of the main pioneers the original telescope. In those exciting days a few hundred years ago, the people who pushed the frontiers of theory also pushed the frontiers of experiment.

Nowadays it's different, you can't always build them at home. They're big and expensive, sometimes noisy too. But whether small-scale or large-scale, experiments reveal a lot, and nowadays they're very good. And experiment is still the final arbiter in physics. So if one believes PSG may be right, there's a need to get experimental evidence that can show what's really going on, and distinguish between PSG and standard theory.

As always, experimenters and phenomenologists will be far better at finding ideas for creating experiments. But I can set out some underlying principles that might provide a few targets, and later I'll describe some mathematically specific experiments.

103. Ideas on quantum mechanics

I'll start briefly with quantum mechanics, or QM, which was in the first book. This chapter has a summary (more complete than the one on page 39), and a few points to add, about aiming for experiments.

The part of the background theory that covers this is dimensional quantum mechanics, or DQM. It's an interpretation for QM, and a particular version of it - the relational approach. Relational interpretations started with Kochen's 1979 version. DQM explains in a new way four phenomena of QM, which are not explained anywhere else in this way, including in Rovelli's RQM, where to be honest not much is explained. They are: superposition, the wave-particle duality, quantisation, and the 'scale issue' - matter's very different behaviour at two different scales in QM. I'll summarise them here.

Superposition. The dimensions have been taken literally for a few decades. If one does take them literally, they're *the only other thing we know of*, apart from a quantum wave, that can be in a physical superposition of possibilities. That's because the dimensions are set at many angles at once - orientations - so there's a set of possibilities about where the axes are in 3-space. In DQM these two kinds of superposition are linked, and one explains the other.

In DQM, as in some string theories, the dimensions make parallel cylinders at the Planck scale. It's a single structure, but how it's positioned is undecided. Light and matter are vibrations in the fabric of this structure, that travel in certain directions through it. So until a local positioning for the axes has been established, not only are the dimensions in an unfixed state, light and matter are as well. It creates a set of possibilities about light and matter.

The wave-particle duality. *The wave state:* when the axes have no defined positions, light and matter, when travelling, make waves. An emitted photon spreads out into a wave, as it takes many paths at once - each is a possible orientation for the axis it travels on. The many versions of a particle make a 'smeared out' object, with multiple versions superimposed.

The particle state: when a local positioning for the dimensions is established, one single orientation is picked out from many, leading to particle behaviour. Matter is now in its well-defined state, and can have a definite location.

Quantisation. A quantum wave is the same particle seen many times. This explains why the wave's energy can only be divided into fixed, equal units. Elements of the wave are different versions of the same particle, so they'll all have the same energy. This energy unit can go no smaller within a particular wave, but as we know, in a different wave it can be smaller - this is because a different wave is based on a different particle.

The 'scale issue'. Matter behaves differently at two different scales because at a large scale, such as in the lab, there's already a network of links between matter. It arises from entanglements caused by interactions. This idea is well supported: decoherence has shown clearly, via both theory and experiment, that when new matter interacts with laboratory matter, entanglements are rapidly created, and the possibilities in the wave are reduced.

According to DQM, there's a network of entanglements in the laboratory. It's relational only, and is emergent. It's like a framework, which allows matter to join onto it. This happens by matter interacting with - colliding with - other matter, already joined to the network. Matter gets 'relational information' (or 'get its bearings'), via an implied positioning for the dimensions.

When a measurement is made, there must be a collision, which explains why this process can be set off by a measurement. In decoherence, we know a rapid series of interactions creates entanglements, reducing the wave state. DQM says that matter is joining onto the network in the lab. The wave state breaks down, the network builds up. Matter takes on the particle state, in which there is a fixed positioning for the axes of the dimensions.

I hope that summary makes sense. Returning to looking for possible ideas for experiments, or principles that might lead to them, perhaps these networks can be shown to exist somehow. When they're created, one thing that gets defined is the reference frame. At present the change to particles is thought to be unconnected to frames. So however it's done, one possible aim would be to look for a connection between them.

But the key place might not be state reduction, but entanglement. In a 2003 paper, a team working on quantum technology at MIT said:

While quantum entanglement as a resource has been studied extensively within the last decade, it was realized only recently that this resource is frame-dependent, and changes non-trivially under Lorentz transformations. In particular, Gingrich and Adami showed that the entanglement between the spins of a pair of massive spin-1/2 particles depends on the reference frame, and can either decrease or increase depending on the wave-function of the pair. A consequence of this finding is that the entanglement resource could be manipulated by applying frame changes.

So a link with frames is indeed found there. Quantum entanglement depends on the frame. As well as giving support to frame-dependent interpretations such as both RQM and DQM, this means that there might be a path towards an experiment. The MIT paper points out that it's possible to manipulate the entanglement by applying frame changes.

The Frauchiger-Renner thought experiment of 2018, which was confirmed in the lab via a similar setup in 2019, gave strong support to frame-dependent interpretations, such as RQM and DQM. That thinking came from the original quantum theory, as did Bell's theorem, so occasionally what we already have can lead to experiments with bearing on the interpretation side.

There are other ideas that might lead to experiments, some of which came after Book I was published, out of long conversations with another physicist who became interested in my theory. Mike Parker, who's a visiting Fellow at Essex University, found DQM interesting after reading the Sunday Telegraph article *'Quantum mechanics' greatest puzzle 'solved' in a Surrey cottage'*. He contacted me, we met up in a pub in Surrey and talked for three hours, also a lot by email. Those conversations made me think hard about experiments that might support my interpretation for QM, and led to some ideas.

Although experiments are definitely needed, in reality, one loose prediction is that the DQM interpretation should become more or less unavoidable with further work. That means - as in the kind of interpretation Matt Leifer says we need to find - indispensable in explaining QM. The wave function should continue to be describable via a set of positions for the dimensions. And QM should be rederivable from that basis, and should fit this picture like a glove, in many places.

I believe that will happen, but having spent twenty five years on the theory, I decided to publish it without delaying yet again. It made good sense to try to get some help (when you call in the cavalry, you hope they won't steal your valuables). But physics is about teamwork, and it seems likely that the work on QM - getting the mathematics back out from the concepts - will be done sooner or later by myself, others, or both.

So the QM part of the theory was published as it was, without developing the mathematics. DQM came late in the process: the background theory was finished by 2005, the core of the gravity theory by 2011. But DQM was only just in shape when I went to see Carlo Rovelli in Cassis to talk about QM in 2017. The mathematics is not in that area of the theory, it's elsewhere. And as for the experimental side, with QM interpretations generally, experiments are extremely rare. So it may be difficult to tell DQM from other views - but there are still ideas, and it might turn out to be possible.

104. Looking for clarity

Returning to gravity, there's a need to find a way to distinguish between GR and PSG. There are places where they diverge: I've mentioned a few. One is black holes, which should have matter sitting near them, and there are very strong hints that this is showing up in the data. But although I see this as exciting evidence for PSG, it's limited - the area very near a black hole is not one of the places where PSG has a detailed description. But a little further out the description is specific, and the numbers are unique.

That's because it's about orbits. And with orbits in general PSG is specific, the equations are exact, what they describe is well understood. The differences from standard theory are sometimes measurable. With elliptical orbits, the more eccentric (and so radial) the orbit, the more the differences will show up. Hyperbolic orbits are even better, as they're more radial. One problem is that solar system orbit parameters are interdependent, and hard to isolate: a list of interlinked numbers might all be slightly different from the estimated values, so there'd be a need to create a PSG ephemeris.

What about deriving the central mass value from a specific orbit? For orbits near to circular, the speeds are almost identical in GR and PSG, so the mass value is too. But calculating it from different heights leads to a test of each theory's consistency. Circular orbits don't work for this, but it can be done via radial orbits - objects in vertical freefall (Chapter 109). Of all the experiments outlined here, this is the best one. That's because what both theories predict is clear cut. And it potentially tests one of the central long-standing ideas in PSG, rather than any peripheral additions.

In the inner solar system, Mercury's orbit may hold a few surprises. It's quite eccentric, and is in comparatively strong gravity. And there have indeed been some anomalies with Mercury. There's also new data from a planet orbiter - it has resolved some questions, but has opened up others.

Apart from orbital speeds, there are other differences of that kind, but most are too small to measure. The equation for the geodetic effect was good for showing an alternative interpretation, but it only gives unique numbers after eight or nine decimal places, so you can't tell them apart. That's measurable only in principle - perhaps until the next generation of instruments.

Looking for clarity, there's a need to get to a numerical prediction from PSG, that can be pinned down in an experiment. And a possible way to get to that exists. Like the installations of LIGO, it involves an interferometer.

It can be a straightforward Michelson interferometer, on the ground rather than in space, like other similar experiments. But there's a difference - one of the arms is vertical in the gravity field. During the hundred years it took to go from Michelson-Morely to LIGO, interferometers have often been L-shaped, but they're almost always laid flat on the ground. This makes the experiment easier to do.

But it's still possible to build one on Earth with a vertical arm. There are also two space interferometers being developed, in Europe and Japan. If they're launched, they'll be far more sensitive than ground-based ones. And they'll move around in a way that behaves like a vertical arm interferometer, when they're in certain positions in space.

The question of what that kind of experiment would do is closely related to the time delay of light experiments. And interestingly, among the published papers on vertical arm interferometers, there has been disagreement about exactly what GR says would happen.

But this kind of experiment can potentially differentiate between two very general principles for gravity: curved spacetime or refractive medium. More specifically, it might show which theory, GR or PSG, describes a gravity field better. So it might reveal which of these two 'on paper' versions of the world fits better with the actual, clunky, out there, real world.

105. A measurable difference

So now we get to a place where the two theories stop agreeing. Or rather, to where their differences might be coaxed out into the open. The time delay of light - the Shapiro time delay - has a kind of quirk in it, which means that a refractive medium behaves differently from curved space. In the Sun's field, inside Earth's orbit (in relation to an Earth clock), GR and PSG give the same results. But outside Earth's orbit, they give different results.

In GR there are various ways to define things. The theory has been worked through for a century in fine detail, so there can be a lot to take into account. What I'll say now won't go far into them, and the literature is ambiguous on this question. So I hope to get the GR prediction from one of the well-known relativists, such as Clifford Will or John Baez. But I'll set out here what PSG does in this area, and it seems different from what GR does there - as GR has an effective cancellation, related to the time delay of light.

This cancellation isn't well known (though Clifford Will has mentioned it) and it doesn't necessarily show up in the standard approach to GR. It shows up in a particular area, if one is timing distant light signals using an Earth clock. The two things that cancel, the space and time components of the Shapiro delay (on radial paths, outside 1 AU), can be in separate areas of the calculation. And the standard use of GR avoids this kind of approach, for good reason. It's questionable to relate distant events to a clock on Earth. But in the Shapiro time delay we do exactly that, so this is an exception.

Imagine the Sun's field, with local clocks dotted around here and there. The further out you go, the faster the clocks run, in both GR and PSG. Inside 1 AU so nearer the Sun, they run slower than an Earth clock. Outside 1 AU, further from the Sun, they run faster than an Earth clock.

Each distance from the Sun has its own time rate. But everything I'll say now is in relation to our Earth time rate. There's nothing special about a radius of 1 AU, except that it's where we keep most of our clocks. And for simplicity, this is always about signals that travel radially in the field.

One point that might help with understanding this is that in GR, in relation to an Earth clock, the speed of light is not constant - it varies. It's only constant in relation to a local clock, ie. a clock that's in the same place as the light. But no clock is ever in the same place as a beam of light for long. In practice, in a place like the solar system, light quickly travels off into other regions, where time runs at different speeds. So to time light so that it actually *behaves* as if it has a constant speed, one needs to use a changing time rate. But no-one does that, as it would mean some very cumbersome calculations, for all the clocks the light would zoom past on the way.

One can put it however one chooses, but it's far simpler to time light as we time more or less everything else: by an Earth surface clock. So that's what people often do, but if you do that, its speed varies.

According to GR - and also PSG - time runs slower inside 1 AU, faster outside 1 AU. The bottom line of that is simply that if you're timing light signals from Earth, it's exactly as if the light travels slower inside 1 AU, and faster outside 1 AU. This isn't the only way of seeing it, but it helps when trying to visualise what follows.

It's easiest to start with inside 1 AU. Inside the Earth's orbit, the two theories agree. There are two parts of the Shapiro time delay: one due to space, the other due to time. These two parts of the time delay have been described by Clifford Will in his well known book 'Was Einstein right?'. They're equal, and their causes in GR are space curvature and time dilation. Their causes in PSG are refractive medium and time dilation. Both parts make a positive delay, in both theories, which means that inside 1 AU, the light arrives late.

But outside the Earth's orbit, time runs faster. So the sign flips with the time part of the delay, in both theories. There's a speeding up of the signal, which means a negative delay: the light arrives back *early* when it returns to Earth. On the time part of the delay, again the two theories agree.

So far, there has been agreement in three out of the four combinations, but on the fourth they diverge. The first three are space inside 1 AU, time inside 1 AU, and time outside 1 AU. But the fourth thing to look at is the space part of the effect, outside 1 AU. And that's where GR and PSG finally shake hands and go their separate ways.

In GR the space part of the Shapiro effect outside the Earth's orbit behaves like a *slowing* of light. The space part always does that in GR, because in any direction in any gravity field, curvature always means the signal has further to travel than the straight line (3D) Euclidian distance. So travelling the extra distance around the curve always delays the signal.

But by contrast, in PSG the space part of the effect outside 1 AU is a speeding up of light. That's because the refractive medium goes on getting thinner as you go outwards, which means light can travel faster further out. So in PSG the two halves of the delay are both positive inside 1 AU, and both negative outside 1 AU. There's a simple symmetry in PSG, because what makes the field is graded in a straightforward way, which means that by an Earth clock everything goes slower inside the Earth's orbit, and faster outside it.

But GR isn't symmetrical in that kind of way. Instead, the space part and the time part cancel each other outside 1 AU, because the time part flips over to a negative delay. So inside 1 AU both halves of the delay are positive, outside 1 AU, one is negative, the other is positive. So being equal, they cancel.

Now this is an oversimplification, and as I said, the cancellation isn't obvious. There's a need to be very specific in GR, for good reason. There are different coordinate systems, and transformations between frames and time rates: for instance, coordinate time can be used, which is the time rate as it is outside the field, an infinite distance away. And in that time rate, GR and PSG are *exactly* the same in relation to this question.

So for a few reasons, the cancellation isn't clear cut. But as Clifford Will once put it about an aspect of the Shapiro effect 'you can argue that part is a delay and part is an "anti-delay" '. In one way of taking GR, it turns out that the cancellation happens. To get a delay, the signal must be sent into stronger gravity than where the clock that will time it is kept.

You also get a delay if the signal is *received* from a stronger source of gravity. In current astronomy the Shapiro delay is sometimes calculated, even for gravity waves received on Earth from mergers, and so on. But those are from fields that delay the signal at source.

We only know the numbers well for the Sun's field, and the two theories are the same for signals sent close to the Sun, being timed by a clock further out - as in all experiments that have been done so far. It's very natural to use a strong source of gravity for the Shapiro delay - in solar system experiments, the strong part of the Sun's field is naturally always involved. So we never get to where the theories disagree, and we never find any anomalies.

But to measure this, there may be no need for a long distance experiment. The principle might be testable on Earth. If so, it could distinguish between the theories in a way that's cheaper, easier to do, and where it's far easier to see what's happening.

106. Vertical arm interferometers

A Michelson interferometer is a well-known device, so I don't need to give a detailed description of it. It appeared in the second most famous experiment of all time, which helps. The Michelson-Morely experiment might be second to Galileo's, as mentioned earlier, in which legend says he dropped different sized objects from the leaning tower of Pisa.

The principle of an interferometer is that it's easier to *compare* the speeds of two beams of light, than to measure the speed of one. You have an L-shaped pair of tubes (of equal length) along which two light beams travel to mirrors, and return to the vertex, the corner of the 'L'. Where they're recombined a pattern of fringes appears, showing how the waves are mixing. This scans the beams accurately - if the fringe pattern moves slightly, it tells you that either the length of one arm has changed, or equivalently, the speed of the light travelling along one arm has changed. And because the light's wavelength is known accurately, the fringe patterns lead to numbers.

Modern versions are more sophisticated, and there are other devices, such as particle interferometers. But to illustrate the principle simply, I'll describe this in terms of the original version. Although an experiment like LIGO is far more advanced than the first interferometer, they're the same in principle, and they're still usually built with both arms horizontal.

If one arm was vertical, that's a different experiment. Instead of looking for ripples coming from elsewhere, it looks at the Earth's gravity field. It's harder to do, but the technological challenge is minor next to some we surmount nowadays. What it does is to compare the two basic directions in the field. And this variation on a well known experiment is closely related to the time delay of light experiment.

None of the experts on GR have mentioned vertical arm interferometers, as far as I know. Clifford Will is about the world's top relativist, partly from his contribution to the parameterised post-Newtonian formalism of GR in the 1970s. What's more, he has made the tests of GR his special subject. As well as writing several books on it, he has set out what GR predicts for a list of experiments in a single document, which has become an excellent reference source: 'The Confrontation between General Relativity and Experiment'. It doesn't mention vertical arm interferometer experiments, but Will could be depended on to give the correct prediction from GR, though other relativists certainly could as well.

But as things are at present, there's no official prediction to refer to for what such an experiment would do, and in principle at least, for all we know, GR might predict that the interferometer would explode.

That seems unlikely, but the point about this experiment is that it tests not just between GR and PSG, but generally between curved space and refractive medium gravity. So it's more important than it might be. And some relativists will surely be aware that if GR was wrong, then RM gravity might *somehow* eventually turn out to be the underlying picture - before the 22nd century. That mention of RM gravity by the Anderson team at NASA looked almost like an insurance policy, mentioning it just in case future generations look back knowing they were wrong.

That may or may not be so, but anyway, this kind of experiment is important. As far as I know, other proponents of RM gravity don't seem to have noticed it. But what it does is, it tests between two general principles - gravity caused by curved spacetime, and gravity caused by a refractive medium.

The literature, up to now, is unclear on what GR predicts for a vertical arm interferometer. In the papers I've read on the subject, I found three different answers to the question, though the exact question varied between papers. One gave a null result, two others gave different numbers for the difference in light return time between the vertical and horizontal arms.

Interestingly, there was also an attempt actually to build such an experiment in Peking in 1990, but the result was inconclusive, and the experiment was never completed. Just to be clear, I'm not saying that one exploded, and I'm not going to go into the question of what GR says would happen, or what the individual papers on it say. What I will do though, is to set out what PSG predicts would happen in such an experiment. There has been no consensus on the GR prediction - nothing from a recognised source as far as I know, and disagreement elsewhere. But I'll give the prediction from my end, and see if any proponents of GR will give theirs.

Part 20. Predictions and experiments

107. An interferometer experiment

When it was the Sun's field, the 'time rate radius' was at Earth's orbit, 1 AU. That's just where the experiment's clock is kept. If we're testing the Earth's field, the time rate radius is at the Earth's surface, and the distance is from the centre of the Earth.

So in the simplest version of this, you just have an interferometer with one arm vertical, one horizontal. The corner of the L-shape, the vertex, is where the clock is kept. It's on the ground, and measuring Earth surface time rate. The vertical arm goes slightly outside that radius, further from the centre of the field, so it goes into an area where GR and PSG disagree.

The two arms are of equal length, probably a few meters. The horizontal arm can be taken to be at the same radius all along it (we can ignore the curve of the Earth's surface). But unlike the horizontal arm, the vertical arm extends upwards through a varying field, and conditions change along the arm. This is why the experiment is so interesting - it probes the field as the time delay of light experiment does, but it's far more controllable.

So here's the PSG prediction. I'll start with a short range version, for a small-scale interferometer, in the size range of existing interferometers on Earth. If one arm points radially away from the mass, the light is faster on this vertical arm, and returns to the vertex quicker than the light on the horizontal arm does.

Just as in the Shapiro time delay of light, there are two equal components to the 'delay' (in this case the light arrives early). One is due to space, the other to time. I'm going to allow for both, and express the change to the speed of light, for simplicity, as if it was a change to the length of the vertical arm. This is easier to set out clearly.

The difference to the speed behaves like a difference in length Δd between the two arms, the equivalent of a shortening of the radial arm by

$$\Delta d \approx GM(d/rc)^2 \tag{47}$$

where:

d is the length of the radial arm
M is the central mass
r is the distance from the centre of the mass to the radial arm's midpoint.

Both arms might initially be built laying horizontal, and made of equal length. One possible way to build it involves making the apparatus rotatable, so that the interferometer can start with both arms flat. The apparatus is rotated in a series of steps up towards a vertical position, one arm staying horizontal. The angles on the way should show a steady change according to PSG, but there's probably a better way to do it.

The equation above applies when the moveable arm is in the radial position. But when it's at an elevation angle θ between 0 and 90 degrees, the effect is equivalent to a shortening of the arm by

$$\Delta d \approx GM(d/rc)^2 \sin \theta \tag{48}$$

It's simply multiplied by the sine of the elevation angle. In some cases, such as with a space interferometer (several space probes working in unison, and sending lasers to each other), there should be two values for r, one for each end of the arm. Earth-based interferometers have comparatively short arms, so equation 47 works well for them.

But where the long range equation must be used, $r_{[1]}$ and $r_{[2]}$ represent the distances from the mass to each end of the vertical arm. Keeping to the L-shaped format for simplicity, the effective difference between the horizontal and radial armlength is then:

$$\Delta d \approx (GM/c^2) \, (\ln [r_{[2]}/r_{[1]}])^2 \tag{49}$$

where

$r_{[1]}$ is the distance from the mass to the vertex of the interferometer
$r_{[2]}$ is the distance from the mass to the end of the radial arm
GM is the central mass.

This gives the right numbers, whether the radial arm points in or outwards from the vertex. (When pointing inwards, the change is then equivalent to a lengthening, not a shortening of the arm.) Although an approximation, this equation is very accurate, and works at any range.

When the radial arm is instead at an angle θ to the field, as it would be for instance with the LISA space interferometer, the apparent shortening of the arm is given by

$$\Delta d \approx (GM/c^2)(\ln [r_{[2]}/r_{[1]}])^2 / \sin \theta \qquad (50)$$

or, using radii instead of the angle, by

$$\Delta d \approx (GM/c^2)(\ln [r_{[2]}/r_{[1]}])^2 / (d/[r_{[2]}/r_{[1]}]) \qquad (51)$$

These give the apparent difference in armlength. But what really changes in PSG is the speed of light.

LISA is still on the list for the European Space Agency, although NASA have pulled out. The system spans distances in millions of kilometers. It involves three space probes flying in a triangle far larger than the Earth, following the Earth in its orbit around the Sun, and connected up with lasers. LISA is like a huge sail dragged behind the Earth, turning slowly as we go around, as if we were flying some giant kite. If LISA is launched, although the aim is not to do that kind of experiment at all, but to find gravitational waves, the differences between GR and PSG should be clearly visible in the data.

108. Clues from existing data

Returning to the LIGO interferometer, there's a possible source of evidence to be found in data that has already been captured and stored. With existing data, unless you're looking for something in particular, you're not necessarily going to find it. In the past, existing data has turned out to contain more that was initially found in it, and people have gone back to old data and searched through it in entirely new ways. A good example is the reanalysis of the data taken by the Mars landers in the '70s.

LIGO's arms are each 4 km long. They're both horizontal in the Earth's field at all times. It took me ages to realise it, and it's a lateral jump, but they're not horizontal in the Sun's field. And although LIGO wasn't intended to explore the Sun's gravity field, it might perhaps be used to explore it.

In the Sun's field the arms are constantly turning and shifting, as the Earth's daily rotation swings them around through different angles. Sometimes the plane of the 'L' will be tangential to the Sun, sometimes one arm will be vertical, one horizontal. Most of the time they'll be somewhere in between. The fact that their positions are constantly varying might provide a shortcut to some information that they were not built to find at all.

Even at 150 million kilometers, the Sun's gravity is strong where we are. And because LIGO changes position in that field, the system should detect a small steadily shifting signal, like a sine wave, moving between the equivalent of a difference in armlength of zero, and a difference of around 10^{-12} meters. And the signal should have a diurnal timeframe to it, so it should go through one cycle in 24 hours. LIGO is easily sensitive enough to register it.

Unfortunately the tides do something similar, and have an effect on LIGO - at a much larger scale. The true armlength is actually changed slightly when the tide sweeps in, because the Earth's shape changes a little, which stretches its crust. Whether the weaker diurnal signal can be extracted from the stronger one is not clear at all, but it's not impossible. Over time the tides shift their rhythm, and the gravity signal would do the same, but the two signals would shift differently. So perhaps the quiet little PSG signal might be found.

I realised in about 2004 that although LIGO was built to detect gravity waves, it might also be used to test PSG in this kind of way. In 2007 I published the prediction in a peer reviewed journal, mentioning the idea of looking to see what the Sun's field does to LIGO. A calculation had led to the figure of 10^{-12} meters, which was right, in the sense that it's given by the above equations for interferometer experiments, using the Sun's GM value, and 4000 m. But at the time I had only the calculation, and not the equations.

GR on the other hand, because of the cancellation I've mentioned, seems to give a null result when one arm points outwards - no diurnal signal. But there might be a difference when one arm points inwards towards the Sun. When LIGO is doing that, GR and PSG tend to agree, and the setup then loosely resembles the Shapiro time delay tests that have been done.

Although the GR prediction on this kind of experiment is unclear, it would be generally good to establish it, as among other things it's a possible test of GR. But although in the case of other interferometer experiments the idea can work, with LIGO and the Sun's field, it may not. The data that could hold the information, although it almost certainly exists somewhere, may simply have been lost in the tide.

109. Identifying a pattern: calculating the Earth's GM

The second experiment has an interesting thing about it. As well as showing PSG to be right, it could also show standard theory to be wrong. Because of that, it's an important prediction in the context of PSG, and one of the best. Again, it's about finding a way to check whether gravity is caused by curved space, or vibrating space.

Standard gravity theory is usually a mixture of Newton's work and Einstein's, but sometimes Leibniz's work comes in as well. He's less of a central figure generally - he was largely self-taught, but highly influential in his day, and for centuries afterwards. His gravity work eventually went into obscurity, except for one equation. The vis viva equation is very important, and it covers all the different kinds of orbits. NASA uses it every day to navigate the solar system.

It's the only part of Leibnitz's work that has made it into what's now standard theory, so I'm sorry that I'll be trying to show it was wrong! But whether or not my own approach is right, that's just how it goes in physics: every era has equations that will be replaced by different ones in the next era. This is only possible because physics is full of equivalence.

The Earth's GM is known far more accurately than G or its M. We're less sure how it separates into two numbers, but we know exactly what you get when they're multiplied together. So GM is good for this experiment.

Now of the two basic directions in a gravity field, the orbital direction doesn't help. There's almost no difference between PSG and standard theory there. If you calculate the Earth's GM from a given circular orbit around it, PSG and GR will give the same number. Go to a different radius, and they're still self-consistent: all four calculations will give the same number. The current best estimate for the Earth's GM comes via the circular orbit of one of the Lageos satellites, with GM = 398600441800000. The PSG equivalent number is very similar, around 398600441800000.0001.

But if an object is moving radially - say, falling vertically, differences appear. And it's comparatively easy to measure a falling object's speed, and calculate the Earth's GM from it. Different theories will then give different values for the Earth's GM, from the same measurement.

And measuring speeds of falling objects, and calculating the Earth's GM from the result, there's a way to narrow things down. These measurements should reveal an inconsistency in standard theory that's just waiting there, ready to be found, if PSG is correct. It should be that if you calculate the Earth's GM from a measurement at one height (using standard theory), it gives a slightly different number than calculating it from another height. The nearer you go to the Earth, the higher the value for the Earth's GM you'll arrive at, if it's via a measurement and standard theory. But it's a very small difference.

By contrast, working from this kind of measurement, PSG should give exactly the same value for the Earth's GM, from any height. This PSG value for GM will always be lower than the number from standard theory. And it's possible to translate between the two, and show a simple pattern. Standard theory, at any height, should give the PSG value for GM multiplied by

$$1 + (2GM/rc^2) \tag{52}$$

That's a small difference, but it's an interesting one, because the result varies depending on where the measurement is made (because of the radius term). There's plenty of technology that could be used to test this, but I'll start by describing the experiment in its simplest form, to show the principle. In fact, as always, it would probably be done differently.

In the experiment, you hold an object still in the Earth's field (any field would do in principle), and then let it go. Remove all other forces like air resistance, so put it in a vacuum shaft. When it has fallen a certain distance you measure its speed, near the end of its flight. You then look at two equations, one from each theory, that give its speed - and try to decide which equation it's going by. What you find is that it could be either.

But although it could be either, the two equations will imply different values for the Earth's GM. Both will be closely consistent with the object's speed at the end of its trajectory.

Then the experiment needs to be repeated at a greater height, using a plane. To boil the experiment down to its essence, it's two velocity measurements at different heights. Airliners go up to about 45,000 feet. Hafele and Keating took atomic clocks on commercial jets, but that doesn't mean this is easy to do. The second run should be done at the same map location as the first, so the Earth's field is the same for both. If a mountain is used, there'd be a need to consider the extra gravity of the mountain, so doing it on a plane is better. You can't use the ISS for this, as it's in free fall itself.

So the second time, you do the experiment at a much higher height. And this time, standard theory should give a slightly different value for the Earth's GM than it did on the ground. But PSG should give the same value.

Equations 12 and 13, which are on pages 103-4, can be adapted for this. Both give the speed of a falling object, if you know its speed at an earlier point on its trajectory. Equation 12 is the PSG free fall equation, 13 is an equivalent of that from standard theory, and the vis viva equation.

In this case, the speed at one of the two points is zero, because the object is held in the field and then released. So in both equations, you make v_1 equal zero, and v_2 you just call v. Both equations then give the speed of the object. One gives the PSG speed, the other gives the speed from standard theory.

The object is released at r_1, and v is the measured speed at r_2. In standard theory:

$$v = \sqrt{2GM(1/r_2 - 1/r_1)} \qquad\qquad\qquad (53)$$

In PSG:

$$v = \sqrt{c^2 - [(c^2 - [2GM/r_2]) / (1 - [2GM/r_1 c^2])]} \qquad (54)$$

That's basic gravity, but there's the option of bringing in Einsteinian gravity, putting in the post-Newtonian adjustments from GR. On radial paths, they're almost exactly the same as the PSG adjustments, and when you put them in both versions, I'm glad to say that the inconsistency is still there. To add the relativistic adjustments approximately, but well enough for this purpose, the right hand side of each equation is multiplied by expression 4:

$$\sqrt{1 - (2GM/rc^2)} \qquad\qquad\qquad (4)$$

That makes both speeds slightly slower. (In GR it's due to curvature, in PSG it's the slowing effect of the refractive medium itself.)

There are two mathematical rules that define the pattern I'm talking about. The first I've mentioned: expression 52 translates between GM values from standard theory and PSG. But there's also another better rule, and although it came from PSG, it leaves PSG out, which makes it a good find. It's a pattern to be found within numbers derived from standard theory, that is, the values for GM derived at different heights, via measurements. So it's an undiluted way to show standard theory to be wrong, if it is.

If *GM* is the value for the Earth's GM derived from standard theory at a given point *r* in the Earth's field, via measuring radial free fall speeds for matter, then GM' (the value derived in the same way at another point r'), will be:

$$GM' \approx GM \left([1 + (2GM/r'c^2)] / [1 + (2GM/rc^2)] \right) \qquad (55)$$

('In the same way' should include the pathlength being the same, though the figures don't change significantly if you vary the pathlength.) This rule should be found to exist within standard theory, and very accurately. And the point is, if it does, then standard theory is, quite simply, wrong.

This leads to a factor difference between derived GM at the Earth's surface and derived GM' in a plane at 45,000 ft, of around 1 + (3e-12). To check this, one can use the vis viva, and the PSG orbital speed equation (equation 26), with or without the relativistic adjustments on both. Or it could be done in a straightforward way, via these equations for GM values. In standard theory:

$$GM = v^2/(2/r_2 - 2/r_1) . \qquad\qquad\qquad (56)$$

In PSG:

$$GM = v^2/(2/r_2 - 2/r_1 + [2v^2/r_1 c^2]) .$$ (57)

Now there's a lot of technology designed to test acceleration due to gravity. This experiment is about speeds rather than acceleration, but in some cases, such as atom interferometry, the devices measure velocities. The technology has been getting increasingly accurate, and this experiment may provide a good clear cut test between the theories. There are two reasons for that. For one thing it's simple - there are not too many moving parts.

But it's better than that. The real point is, it's not just a numerical prediction, it identifies a pattern. And a pattern is far easier to find in the data than just loose numbers. With loose numbers, the result might potentially be caused by something else. But that pattern will either be sitting there in the data or it won't. And whether or not that pattern is found will have direct bearing on the question of what causes gravity.

Carlo Rovelli once asked what experiment would prove my theory, assuming unlimited funding to do it. I thought I'd say, for a joke, that if there really was unlimited funding, we might not bother with the experiment. But I didn't say that, and didn't mention gravity. We were talking about quantum mechanics (he's worked on quantum gravity for a long time, and he was looking a bit worried already). So I couldn't answer that question at the time - to answer it properly, I'd absolutely have to talk about gravity experiments.

But I can answer it here. If funding was unlimited, this experiment might be built into a plane that flies very steadily, with the apparatus protected from vibration. The first test would be done on the runway, the rest at different heights. But it might be easier: nowadays there are portable gravimeters not far from the accuracy needed. I'm not the person to say how it can be done, but I can give examples of possible ways.

The pattern it would be looking for is accurate out to well above the Earth's atmosphere. It's still spot on and accurate out at the distance of the Moon's orbit. I then tried it at 20 times the Earth-Moon distance, where things fall very slowly (this is still the Earth's field, so gravity is weak out there). You find the pattern has become a little less accurate, but not much, and we'd do the experiment near the Earth anyway.

This prediction, for a free fall gravity experiment, has some advantages over all other experiments I've looked at. But there's also the Galileo experiment, which could lead to measurements. PSG gives a slightly different prediction from standard theory for when light and matter, released at right angles, hit the ground. If such an experiment was done, two mirrors with light bouncing between them would probably be the way.

But with the free fall experiment the good thing is: what GR says is clear. And what PSG says is also clear. And it can be done. If PSG was shown to be right, the equation that would then be confirmed originally came out of the helical refraction equation. And that equation, in turn, came not only out of Snell's law, but also out of the mathematics of the background theory, which gives the helical path angle.

So if that unexpected little pattern was found, and if the result turned out to be repeatable, a wide area of the whole theory would be shown to be right, and standard gravity theory would be shown to be wrong. If this happened, a single very simple experiment would tell us a surprisingly large about what's underneath the physical world we live in.

110. Black hole orbits

As in Chapter 99, 'A cancellation', when you derive the special case circular orbit speed equation from the elliptical orbit one in PSG (by making the semi major axis the radius), on the way you find that there's a rough cancellation. It gets increasingly exact as the orbit gets more circular, and less eccentric. If the cancellation was exact, it would reduce to Newton's circular orbit speed, square root (GM/r). But because two terms (that averaged would disappear), are multiplied, it leaves a small residue:

$$v = \sqrt{(GM/r) - 4(GM/r)^3/c^4} \qquad (45)$$

That makes no difference in weak gravity, but in strong gravity, it does make a difference. Black holes are good for holding a magnifying glass to any small differences between theories.

Partly because the Schwartzchild radius is $2GM/c^2$, whatever equation one's using, the orbit speed there, around any non-rotating black hole (whatever its mass) comes out the same. The radius varies, but not the speed. Now the orbit speed at the Schwartzchild radius is hypothetical in some theories, and not real. But it's very real if you talk about multiples of it, like 3 Schwartzchild radii, which in standard theory is the innermost stable circular orbit, or ISCO. That radius was thought to be the inner edge of the accretion disk, as inside that radius, matter should fly into the black hole - but not so.

A Keplerian orbit (often referred to in standard theory), has the ISCO always the same number. Comparing that speed, square root (GM/r), with equation

45, one finds the following orbit speeds for *any* non-rotating black hole. (The first pair of speeds is hypothetical, for reasons coming from both theories.)

The orbit speed at 1 Schwartzchild radius in standard theory is $2^{-\frac{1}{2}}c$, 0.7071c. The orbit speed at 1 Schwartzchild radius in PSG is 0.

The orbit speed at 3 Schwartzchild radii in standard theory is $6^{-\frac{1}{2}}c$, or 0.4082c. The orbit speed at 3 Schwartzchild radii in PSG is $6.75^{-\frac{1}{2}}c$, or 0.3849c.

The orbit speed at 4 Schwartzchild radii in standard theory is $8^{-\frac{1}{2}}c$, or 0.3535c. The orbit speed at 4 Schwartzchild radii in PSG is $8.5333333^{-\frac{1}{2}}c$, or 0.3423$c$.

Further out, the two speeds keep getting closer. But moving inwards, these figures show Kepler goes on upwards, but PSG rises, then descends to zero. (Certain things complicate this: the black hole is probably rotating. Standard theory has a simple equation for the effect of that, which may be similar to what happens in PSG.) But for now it's worth noting that the orbit speeds we actually measure are low, as PSG predicts. One paper says:

'In the disk-like accretion flows, the angular momentum of the matter in the disk is sub-Keplerian everywhere, except the strong-gravity region $r_{pot} > r > r_{cen}$'

That paper was named '*Leaving the innermost stable circular orbit: the inner edge of a black-hole accretion disk at various luminosities*'. Using the word '*Leaving*' implies that some ideas on black holes seem to have been wrong. As you know, in PSG matter can remain in an unexpected place, right near a black hole. As you can see from the following quote, they're finding matter beyond the edge, where it's not meant to be able to go. In the abstract of that paper, the authors say:

'*Thus, in this case, one may rightly consider ISCO [the innermost stable circular orbit] as the unique inner edge of the black hole accretion disk. However, even at moderate luminosities, there is no such unique inner edge because differently defined edges are located at different places. Several of them are significantly closer to the black hole than ISCO. These differences grow with the increasing luminosity. For nearly Eddington luminosities, they are so huge that the notion of the inner edge loses all practical significance.*'

So there goes the borderline, at 3 Schwartzchild radii. But it's an inevitable result of general relativity, and that whole view of gravity.

Part 21. Physics is changing

111. Judging a physicist on their work

As most people know these days, on the internet the truth is only one of the services available. There are untruths and spin. The truth can often be found, but sometimes only by doing some research. It can help to know the general direction of the spin, and in physics, one angle is 'we know what's going on'. We do, of course, know what's going on to a greater or lesser extent in many areas, but one kind of spin exaggerates the extent.

Carlo Rovelli pointed this out in a funny way in the online documentary 'The Interactions Avenue', which has him and myself talking. At one point he said in quantum mechanics, people are divided into camps about the mystery. He listed some different views of it - each part began: "and then there's another group, who say, *'oh* **we** *know what's going on….'*". It's helpful to be aware of this side of physics behaviour, as it gets into some articles, which exaggerate our understanding in a misleading way. But it's one pattern among many.

I've talked about how important the conceptual side of physics is, and how it may be part of the ultimate goal, as Wheeler and Einstein thought. I also said that the credit for a theory such as the one set out here can't go to anyone who takes the originator's conceptual basis, and from it produces a more complete version of the mathematics.

People always extend the mathematics of a discovery afterwards. That's just what happens, and it happens with every breakthrough. But the discovery itself was still made by the discoverer.

Lateral conceptual solutions, once seen, can seem obvious, almost as if they were known all along. Benôit Mandelbrot, whose life and work my publisher Nigel Lesmoir-Gordon documented in three films and a book, used to say something about that. He said when he discovered fractals, at first he was ignored. Later people took it as correct, but irrelevant. The final stage was: people would imply it was important, but that they knew about it all along. Physics can also be like that. A truly new idea, if right, may be ignored, then resisted, then later accepted - sometimes with the implication that it was always in our vocabulary anyway.

There's also an important point to make about journals: some journals are perfectly respectable, but some describe them as 'disreputable', who quite simply *disagree with the views they publish*. There's a correlation between how disreputable a journal may be described as, and how much it publishes ideas that go against the mainstream view.

So although some journals genuinely publish badly researched papers, there are also low profile, left-field journals that publish good work from physicists around the world, many of them in universities. But that kind of journal may very naturally have a low impact factor - that's a number based on how often its papers are cited in other journals. A low impact factor, at times more or less by definition, is closely related to being *outside the mainstream*. But a low impact factor is seen by some as being 'disreputable'.

So again, being outside the mainstream can be linked to disreputability, even though the mainstream view may be right or wrong. To put it in perspective, so far in the history of science, the mainstream view has turned out to be wrong every time - without exception. So science's task is to look at both possibilities. Putting down work that opposes that view at times contains the assumption that the mainstream view is right - and that's bad science.

But physics often get entrenched in a standard view, and so is less able to make progress. It's very much easier to get funding for work that reinforces the standard view, and riskier to ones career to question it. So you get a central edifice, and it keeps getting reinforced, even if it's wrong. People tend to focus on where it agrees with experiment, and turn a blind eye when experiment contradicts it. So physics is sometimes like one of those toy cars that can only go forward. If it drives into a cul-de-sac, as it may have done with GR, it can't easily reverse out again.

Alongside that, there are many who have put decades of work into standard theories such as GR. There's no conspiracy, of course. But some are naturally reluctant to see their work suddenly become irrelevant, and they sometimes try to disqualify new ideas for that kind of reason.

But this century things are changing. Thomas Kuhn has already shown that the history of science is a series of revolutions, in the most influential book of that kind ever written, 'The Structure of Scientific Revolutions'. Many now recognise that people like Einstein, who was an outsider - being an amateur without any academic affiliation when he published special relativity - have provided the ideas for such revolutions, and will often find better ideas than those nearer the centre. To think outside the box, it can help to be outside it. So the list of outsiders includes Galileo, Leibnitz, Faraday, in biology Darwin, and many others. And those are the ones we've heard about, which is only the tip of the iceberg. They were in a small group who were lucky enough to publish at all.

But nowadays there's a new recognition that a physicist should be judged by his or her work, and *not by anything else at all*. That includes the other work it gets published alongside. As long as the journal has a genuine peer review process, nothing else makes a difference. And nothing of that kind should be useable in the squabbling that sometimes goes on between physicists, and in their attempts to disqualify each other.

But although things are changing, some have not changed enough. Outsiders made many of the great breakthroughs in the past. Ironically though, being an outsider in the 21st century, is still a bit like being a black physicist in the 20th century, or a woman physicist in the 19th. You have to find your friends among the fast growing number of forward thinking people who have come into the field nowadays.

But judging physicists by their work alone is not a new idea. People who do that have always been there, in fewer numbers. That's how outsiders such as women physicists got their papers through at all. When the system was even less fair than it is now, there were always people with values like that, who stood up for those who were discriminated against.

112. The crunching sound of jigsaw pieces

This book has a lot in it, thank you for bearing with me. I'm hoping that all of it together is enough to show GR to be wrong, and something very like PSG to be right. I think to those with an open mind, it shows that this is the case. But in some areas there's enormous prejudice in favour of GR, and not just among those whose funding or position is helped by the apparent success of established physics. Science journalism, at times, favours the standard view more than the physicists themselves - the disparity between what the two groups say about general relativity is sometimes a good example. Journalists sometimes say that new evidence 'shows Einstein was right', while physicists know it's far less clear cut than that, although it makes a readable story. And online, misleading oversimplifications can be found in many places.

There's still work to be done on the theory. Some areas of it are ongoing, and I'm hoping to collaborate on it, and work closely with other physicists. To me this is an exciting possibility, and it's something I've tried to make happen at several stages on the way. What's known as brainstorming, 'knocking ideas around', can be very positive, and I'm sure it would have sped up the slow processes I went through.

But I don't believe in speculating far - I like things you can pin down. I did my speculating in the '80s and '90s, and since then have been following a trail of clues that came out of one of the ideas, trying to pin them down. To me the most interesting side of this book is where things do get pinned down.

There are also places where the ideas cross-corroborate each other in a close network of concepts, and can answer several unanswered questions at the same time. But that on its own isn't enough. The best part so far is the places where to my amazement and delight, the concepts worked well when tested mathematically.

But however good the evidence created that way, that isn't enough either. The ideas have to be tested by experiment, and when they are, they have to fit well with the real, out there world. And I hope they do. Although it's too early to tell, there are good signs from areas including black holes, the flyby anomaly and the missing mass, that seem to support the theory, and so also to give support to the general picture I've been describing. Time will tell, and over the next few years, hopefully from specific experiments, the data that's needed will appear.

People sometimes ask about possible future technology - any idea where the theory might lead? One thing I can mention is the technological revolution that's going on now all over the world, in quantum technology. Not having an interpretation for quantum mechanics hasn't stopped the technology from making good progress in certain areas, but that doesn't mean a conceptual breakthrough wouldn't help. A new understanding of QM could make a huge difference, so I'm hoping the DQM interpretation, which was published in 2019, and recorded in the documentary, will filter through to the people who are connecting the interpretation side to other areas.

And because of the group side of physics, if this picture turns out to be right, or a good description, after a while something I mentioned in Book I should begin to happen as well. Gravity and quantum theory are connected into so many parts of our picture, that people in all different areas of physics should start to find jigsaw pieces slotting into the landscape in their particular field, in new ways. That would certainly include areas I never even thought about at all. So after a while we might start to hear the general crunching sound of many jigsaw pieces all clicking into place at the same time - I don't know if that'll happen, but it'd be good if it did.

THE END

Appendix A: An approximation

PSG has a factor for the slowing of three things in a gravity field: light, matter and, in a different way, time. It's a fixed number at a given point in the field, expression 4, which depends on the radius.

That bit of mathematics was first found by Einstein in a different context. In GR it gives the local time rate and the gravitational redshift. In PSG, as well as those, it gives the local transmission speed of space, showing how everything gets slowed near a mass.

So it's an important number in a gravity field, but there's also an alternative expression, which is very similar. It gives almost exactly the same numbers in weak gravity, such as in the solar system. It's called 'the approximation' in my notes, and I found out some years after finding it that naturally enough, in GR it's also seen as an approximation for the same expression. (Einstein used something very like it early on, to give the varying local speed of light, in his 1911 theory.)

The full expression and the approximation give ridiculously similar numbers, often differing in the solar system by something like 2×10^{-19}. That difference isn't currently measurable, and they're for most practical purposes the same. They really only diverge in strong gravity, and the difference may eventually become relevant in areas such as black holes.

As I've mentioned, if one assumes Einstein's gravitational redshift equation is exactly right, and in both GR and PSG there's reason to think it is, then at any point in the field, in PSG the transmission speed of space in terms of c is

$$\sqrt{1 - (2GM/rc^2)} \tag{4}$$

This is as a fraction of whatever the transmission speed of space would be, at that point in space, if the central mass wasn't there. That's PSG, but in weak gravity it's approximated by

$$1 - (GM/rc^2) \tag{58}$$

This expression closely mimics the other one, because of a general pattern. The square root of a number just below 1 always lands very near to halfway between it and 1. If it's of interest, this can be tried with a calculator. Take a number just below 1, say 0.99999. Find the halfway point between it and 1 (add 1, divide by 2). In this case it's 0.999995. Then, for comparison, take the square root of your original number, in this case 0.9999949999. It's near the halfway point, and the nearer to 1 you start, the closer it lands.

This explains why the two expressions above are so similar in weak gravity - such as in the solar system, where the subtracted term is very small. One of them gives the square root of a number just below 1, and the other takes the halfway point between that same number and 1.

So the approximation also gives the right numbers. If by any chance it turned out to be the correct expression, which is not impossible, the slowing effect of the RM would then *subtract* a speed from the basic transmission speed, rather than slowing it by a factor difference. If so, refraction in general, which may arise (if one looks at a small enough scale) from a series of delays due to collisions, might be better described by subtraction, not multiplication - and I'd have to dig out the alternative set of equations that I wrote down many years ago, which gives more or less identical numbers.

But there's good reason to think Einstein's expression for the time rate is the correct one, coming from quite a few different directions. Anyway, the two expressions are indistinguishable where we live, and for the next few years at least, even our precision instruments can't measure the difference. Until that changes, the other expression is mainly just a way of simplifying certain calculations - physicists and cosmologists use it for making approximations, and it's often very helpful in that way.

Appendix B: The deflection of light and matter

Light and matter have similarities and differences. There's a need to explain both. They're good clues, but standard physics, having less of a conceptual picture that relates light to matter, struggles to explain them in places. But treating light as a particle sometimes works.

In PSG this comes into focus in the Galileo type experiment. The similarities between light and matter are about their Planck scale nature, the differences come from the direction they travel in through the structure of space. With gravity, the similarities arise because they both get refracted, the differences arise because they do that in different ways.

Here I'll show a less than well known fact: the same equation describes the deflection of both light and matter. Einstein's deflection angle for light (such as by the Sun) is usually reached via $4GM/rc^2$ radians. Newton's half of that is $2GM/rc^2$ radians. These work for light, but they simply don't work for matter.

On the face of it, when light or matter pass near a mass, they get deflected in a similar way. The light's path looks like a hyperbolic orbit - an open orbit - but it doesn't change speed as matter on a hyperbolic orbit does. Instead the light zooms past the mass, briefly slowing down very slightly. By contrast, the slow hyperbolic orbits of matter *speed up* on their way in, swing around the mass, then move off again, slowing down as the object drifts away.

These two kinds of behaviour are different, the speed changes are reversed. But if you play with the mathematics we have for matter, and then try it out on light, you can get to equation that works for both.

There are different ways to calculate the deflection angle: trying them out on light gives a few clues. Some of them work for light, via terms that seem very matter-related, and you find that light's 'orbit' seems to have parameters like eccentricity, and a hyperbolic excess velocity. The HEV is the speed matter approaches, but never quite reaches, slowing down as it heads away from the mass. But with light, you put in the speed it always travels at (or the one it speeds up towards on its way out), and the equation still works.

Newton's well known angle for light passing the Sun can be arrived at using equations that many assume apply only to matter. A straightforward version I reached (an approximation, like many ways of getting the angle), looks only at what happens at closest approach. It gives Newton's deflection angle δ for either light or matter, out of three numbers: the mass's μ (GM), the closest approach distance r, and the speed there, v. The deflection angle is

$$\delta = 2\,arcsin\,(1/[(rv^2/GM) - 1])\,. \tag{59}$$

To test this, there are some hyperbolic orbit figures on the Wikipedia page on the flyby anomaly, with deflection angles. Leaving out the Galileo II probe (which has the wrong numbers on that page: two numbers are the same that should be different), all the other flybys work with this equation. You convert km to m, and add altitude to the Earth's radius for the r term.

But if you then put in c for the speed v, and the figures for a light ray grazing the Sun - the Sun's mass and radius - you get 2.4322 x 10^{-4} degrees. That (by 3600) is 0.875 arc seconds, which is the well known Newtonian angle for the deflection of light by the Sun.

This is not meant to be astounding (other parts of the book may be). But it's an interesting point, nonetheless. It means matter particles like neutrinos, which travel very near lightspeed, are deflected in the same way as light. The above equation, if you put in c, won't know which you're asking about, and will give the same result for either. *So there you have a very direct similarity between light and matter, and a mathematical bridge between them.*

The fact that light goes by these equations is likely to be known already, but the similarity is not too well understood. It raises a number of questions. And with the deflection of light, if this gives Newton's half, what about Einstein's? And it becomes clear that in PSG it's not only light that has two parts to its deflection angle, matter has as well. With matter Einstein's part is very small, but Newton's is large. Only at lightspeed are the two halves equal.

So whether large or small, what *is* the Einsteinian part of the deflection? It's just the post-Newtonian adjustments again: either curvature being added on, or the effect of the RM. Everything gets multiplied by expression 4, as per usual.

In the case of orbits for matter, the extra angle of the Einsteinian part only makes a tiny difference, which is hard to detect. But occasionally it comes to the surface, as it did with the perihelion shift of Mercury.

Appendix C: The Gravity Probe B anomalies

I won't speculate about the strange effects the Gravity Probe B team found. Instead I'll quote from a summary by Clifford Will, who was overseeing some aspects of the results, and giving his own view. He's done the same in other situations, being one of the world's top relativists. He wrote:

First, because each rotor is not exactly spherical, its principal axis rotates around its spin axis with a period of several hours, with a fixed angle between the two axes. This is the familiar "polhode" period of a spinning top and, in fact, the team used it as part of their analysis to calibrate the SQUID output. But the polhode period and angle of each rotor actually decreased monotonically with time, implying the presence of some damping mechanism, and this significantly complicated the calibration analysis. In addition, over the course of a day, each rotor was found to make occasional, seemingly random "jumps" in its orientation—some as large as 100 milliarcseconds.

No-one knew for certain what caused these effects. There may have been a reluctance to see them as to do with gravity, and a reluctance to see them as *not* to do with gravity. Either is not too palatable in the situation, as the team spent 30 of the 40 years shielding out all possible effects other than gravity. So measured effects are far more likely to be gravitational effects than usual, and the hope is of course that the shielding worked well.

But to say that these effects *are* to do with gravity would also not have been too good. At least the probe would have done its job, but for gravity physics, it would leave behind some very baffling data. The effects were taken to be 'noise', rather than gravity effects at work, and the team set about removing parts of the data they had spent 40 years trying to get. They also tried hard to get readouts from the gyros, through difficulties, and did well.

If an experiment doesn't do what one expects, one shouldn't start adjusting the data to bring it closer to expectation. It certainly might have been caused by something else, and perhaps one of the things they suggested. But it also may be that PSG can be found in the 'noise' that was removed, so I'm glad that they will have kept the data. It was a highly accurate and beautifully put together experiment. So I hope one day people will look at the data again, and that by then PSG will be on the list of possibilities of which the team will be aware, and I'm hoping they'll keep an eye out for it.

Appendix D: Calculating transmission speeds

There are two ways to think about how light travels through a gravity field. Either is valid. The interpretations in GR and PSG are different, but they're mathematically equivalent, and the results are usually identical.

You can't talk about a speed without deciding what time rate to use, because $v = d/t$. Any speed needs a time rate. But across a gravity field the local time rates vary. And the solar system has many fields, so if you travel through it the time rate changes steadily from place to place. You can either choose to say the time rate varies and light's speed stays the same, or that light's speed varies, and the time rate stays the same.

The more complicated approach in calculations is to make adjustments for clock rates everywhere, and keep a constant speed of light. It's far simpler to use a fixed time rate, as people do when timing things. It's usually the one at the Earth's surface, which we use for more or less everything. The reason physicists don't talk as if light's speed can vary is not general relativity, it's special relativity. That early stage was about matter in the absence of gravity, and it was a key element to say that light has a constant speed - from any viewpoint. That key opened many doors in 1905.

But by 1911 Einstein was taking light's speed to vary, and it's very possible to do that *in relation to gravity*, as he did at the time. He later incorporated the early part of the theory into the later part, and it all hung together well, with a constant speed of light. I'm only mentioning this now as it might help with understanding the alternative, much simpler approach. Some find it hard to imagine light's speed varying, but GR absolutely says it varies if you use Earth surface time rate. This might make what follows easier to understand.

As an aside, although physicists do sometimes talk about a varying speed of light, it's an informal way of talking. Strictly, one has to bring in the definition of the meter: the distance that light covers in 1/299792458 seconds. So if one decided to define things differently (and risk crossing swords with the 'Conférence Générale des Poids et Mesures', the authority that laid down the definition of the meter), there'd be a need to state it carefully.

In PSG there's more reason to have a system using the Earth's time rate for light. Unlike in GR, matter is very like light, so separate systems for light and matter, as people sometimes use in standard physics, isn't the best way in

PSG. The speed of light is the transmission speed of space, and that number is a speed for anything made of waves - and everything is.

So transmission speeds are important. There are different ways to calculate them: the system I've used isn't necessarily the best one. But it was useful when working on the theory. It's like the way we calculate RM speeds for light's behaviour on Earth. In glass, light travels at $0.65c$, and in water $0.75c$. The refractive index is the inverse of the speed.

It's possible to make a similar system for gravity. Then the speed of light can vary, depending on the refractive index of space. Empty space can be 1, and any material more opaque than space has a refractive index above 1. But in PSG, you can't always make the refractive index of space equal to 1, that's the problem. Out in the solar system there are different gravity fields, they overlap, and many fields affect us. We often look at just the Earth's field, or just the Sun's, and then this kind of system works well. But sometimes we need to combine fields.

In PSG that means allowing for the secondary vibrations that are constantly arriving from all directions at a given point in space. Some come from large masses far away, others from small masses nearby. They all slow light down, and their effects are multiplied together - it's similar to 'linearised gravity', which is a simplified version of GR. According to PSG, if I shine a torch across my garden, many fields slow the beam down slightly, because tiny vibrations are arriving there (just beyond the shed) from all directions. So if we want to know how fast the torch beam would travel without them doing that, we'd have to remove all these fields somehow.

It would be good to have a system that starts from a very basic number: the speed of light far from any gravity fields, out in the voids between galaxy clusters. From a viewpoint out there, the galaxies are just far distant places where space is vibrating a lot. (In fact, vibrations from the galaxies that travel out and reach a place like that wouldn't necessarily be negligible.) But in one way or another, it might be possible to calculate the transmission speed of entirely non-vibrating space.

If we knew that number, it could be made equal to 1 in some wider system. But because we don't know it, we have to work backwards from where we find ourselves in the universe, removing one field at a time.

But to do that, a starting point is needed. If there's a point at a radius r in any field at which you already have a value for local c, you can get all the speeds for light across the whole field. We know the speed of light at the surface of the Earth, so that's our starting point. That makes it possible to remove the

Earth's field, and get the speed of light at 1 AU in the Sun's field, but without the Earth there (as, for instance, when it's on the other side of the Sun). It comes out as 299792458.20869 m/sec, about 20 cm/sec faster than with the Earth there, because the Earth slows light down.

And that in turn provides a starting point for the Sun's field. This leads to a system, with everything expressed as we usually express things, by an Earth surface clock. it was just for me to use, there may be a better one. In trying to explain the solar system anomalies, you need a PSG ephemeris. But if you can't get one, calculations via this system can help. For me, it mainly helped with ruling out many ideas.

In fact, trying to calculate the deep space speed of light, working backwards, there can be issues about special relativity, such as how the galaxy is moving, which I won't go into here. But if one allows for that, the calculation can be done: first we'd have to remove the effect of the Earth, then the Sun, our galaxy, Andromeda, the local group, the local supercluster, the huge mass known as the great attractor, and so on. The end result would be an estimate for the deep space speed of light.

I made an estimate for it twenty years ago, in a very approximate, long PSG calculation - for not much reason except that I really wanted to find out that number. It took a day or two, but the result wouldn't necessarily be accurate enough. Since then some figures for the masses are known more accurately, but it still would be very much an approximation.

Although the speed of light varies in some ways of expressing it, in PSG c is still a universal speed. If one uses the local time rate, light will always come out as moving at 299792458 m/sec, anywhere in the universe.

But if we had the deep space speed of light - to an Earth clock - we'd know the speed at which space transmits waves out in the far distant voids, where there's nothing much affecting the waves, and nothing jamming things up, and creating all the complications we find in galaxies. That number might be vaguely interesting, with its fundamental simplicity, but anyone who went all the way out there to investigate it might find things boring after a while. Sooner or later, they'd probably miss the complications, and start to think about turning round and heading for home.

Index

In-depth section

References
References are listed in the format:
Chapter number [page number].

Intro [5] **Raphael Bousso**. *What is your favourite deep, elegant, or beautiful explanation?* Edge.org , https://www.edge.org/response-detail/11707

Intro [6] **Sabine Hossenfelder.** Videos:
https://www.youtube.com/watch?v=PdL8CudJTcs   *[5 mins 19 secs]*
How we know that Einstein's General Relativity can't be quite right.
https://www.youtube.com/watch?v=Ov98y_DCvRY   *[1 min 50 secs]*

Intro [7] **Gravitational waves.** Weisberg, J.M., Taylor, J.H. *Gravitational radiation from an orbiting pulsar.* Gen Relat Gravit **13,** 1–6 (1981).

Intro [7] **Binary pulsar, gravitational waves graph, 1975 to 2005.**
https://www.emis.de/journals/LRG/Articles/lrr-2006-3/fulltext.html
(on the left of the page, click on 5.1, down that page to Fig 7, click on that.)

4 [13] **List of theories with gravitational waves.** Jeffrey S. Hazboun, Manuel Pichardo Marcano, Shane L. Larson, *Limiting alternative theories of gravity using gravitational wave observations across the spectrum*
arXiv:1311.3153v2 **[gr-qc]**

6 [16]. **Rietdijk-Putnam argument** (motion through time). Rietdijk, C.W. (1966) *A Rigorous Proof of Determinism Derived from the Special Theory of Relativity*, Philosophy of Science, 33 (1966) pp. 341–344

Putnam, H. (1967). *Time and Physical Geometry*, Journal of Philosophy 64, (1967) pp. 240–247

6 [17] **Misner, Thorne and Wheeler**, *Gravitation*. 1.3, Weightlessness, p 13. W.H. Freeman, 1973.

11 [22] **Inertial mass, gravitational mass quote.** Einstein, A. *The Meaning of Relativity*. Routledge. p. 59. (2003). ISBN 9781134449798.

11 [22] **Equivalence principle quote.** *On the Relativity Principle and the Conclusions Drawn from It*, Einstein, Albert, Jahrbuch der Radioaktivitat und Electronik 4 (1907), Heading: *Principle of Relativity and Gravitation.*

13 [27] **Discussion site.** https://physics.stackexchange.com/questions/ 410433/geometric-explanation-for-the-equality-of-active-and-passive- gravitational-masse

13 [27] **Active, passive mass experiment.** Kreuzer, L. B. 1968, *Phys. Rev.,* **169**, 1007

13 [27] **Active mass uses G.** Clifford Will, *Was Einstein Right?* Basic Books, New York, 1986, Oxford University Press 1995 edition.

15 [30] **Alternative to Mach's principle.** Alfonso Rueda, Bernhard Haisch, *Inertia as reaction of the vacuum to accelerated motion,* Phys.Lett. A240 (1998) 115-126; arXiv:physics/9802031v1 **[physics.gen-ph]**

16 [31]. **Hořava gravity**. Hořava, Petr, *Quantum gravity at a Lifshitz point*, Phys. Rev. D 79 (8): 084008

16 [31-2]. **Entropic gravity.** Verlinde, Erik P., *On the origin of gravity and the laws of Newton*, JHEP 1104:029,2011, arXiv:1001.0785 **[hep-th]**

17 [33] **Turok, Neil.** Perimeter Inst. public Lecture, *The astonishing simplicity of everything*, Oct 2015. https://www.youtube.com/watch?v=f1x9lgX8GaE

17 [33-4] **Turok, Neil.** Beyond the Big Bang: Searching for Meaning in Contemporary Physics. Part One, The Origins of the Universe: Why Is There Something Rather than Nothing? October 14th 2014, The New York Academy of Sciences, https://www.youtube.com/watch?v=tznxK3etagE (47.07- 49.00).

25 [44] **Anderson quote**. Anderson, J. D., Lau, E. L. & Giampieri, G., 2004, *Measurement of the PPN Parameter Y with Radio Signals from the Cassini Spacecraft at X- and Ka-Bands*, in Proc. of the 22nd Texas Symp. on Rel. Astrophys., Stanford, eConf C041213, 0305

26 [45] **History of refractive medium interpretations of GR.** Saswati Roy, Asoke Kumar Sen, *Study of gravitational deflection of light ray,* 2019 J. Phys.: Conf. Ser. 1330 012002
https://iopscience.iop.org/article/10.1088/1742-6596/1330/1/012002

26 [46] **Dispersion?** John Ellis, N.E. Mavromatos, D.V. Nanopoulos, *Probing a Possible Vacuum Refractive Index with Gamma-Ray Telescopes*, Phys. Lett. B. 674: 83-86, 2009. arXiv:0901.4052v2

29 [52] **Eddington.** Eddington, A. S. *Space, Time and Gravitation: An Outline of the General Relativity Theory* (1920).

33 [55] **Light history quote.** Zubairy M.S. (2016) *A Very Brief History of Light*. In: Al-Amri M., El-Gomati M., Zubairy M. (eds) Optics in Our Time. Springer, Cham. https://doi.org/10.1007/978-3-319-31903-2_1

33 [56] **Moon hammer + feather drop.** Joe Allen, NASA SP-289, Apollo 15 Preliminary Science Report, Summary of Scientific Results, p. 2-11 https://nssdc.gsfc.nasa.gov/planetary/lunar/apollo_15_feather_drop.html https://www.youtube.com/watch?v=KDp1tiUsZw8

42 [68] **Time delay of light.** Irwin I. Shapiro; Gordon H. Pettengill; Michael E. Ash; Melvin L. Stone; et al. (1968). *Fourth Test of General Relativity: Preliminary Results*, Physical Review Letters, **20** (22): 1265–1269.

51 [81] **Test of string theories.** Article: https://www.popularmechanics.com/science/a31898338/nasa-experiment-string-theory   https://chandra.cfa.harvard.edu/photo/2020/perseus https://www.nasa.gov/mission_pages/chandra/images/chandra-data-tests-theory-of-everything.html

55 [87] **Fermi orbiting telescope.** Connaughton, V. et al., *Fermi GBM Observations of LIGO Gravitational Wave event GW150914,* arXiv:1602.03920 [astro-ph.HE]

69 [112] **Non-spherical gravitational fields.** E. Herrera-Sucarrat, P.L. Palmer, R. M. Roberts, *Modeling the Gravitational Potential of a Nonspherical Asteroid*, Journal of Guidance Control and Dynamics 36(3):790-798

71 [117] **Geodetic effect.** de Sitter W, *Mon. Not. R. Astron. Soc.* **77** 155 (1916).

71 [117] **Gravity Probe B**. Everitt, C. W. F. et al., *Gravity Probe B: Final Results of a Space Experiment to Test General Relativity*, Phys. Rev. Lett. 106, 221101 (2011).

73 [123] **Geodetic effect derivation**. Kerr, J., *A derivation of the geodetic effect without space curvature*, Journal of Gravitational Physics, Vol.2, No.2, (2008), received 21.1.'08, accepted 28.2.'08.

78 [132] **Star, source 2 (S2 or S0-2).** Ghez, A. M. et al., *Measuring distance and properties of the Milky Way's central supermassive black hole with stellar orbits*, The Astrophysical Journal, 689:1044Y1062, 2008 December 20.

78 [132] **Sagittarius A*.** Sheperd S. Doeleman et al., *Event-horizon-scale structure in the supermassive black hole candidate at the Galactic Centre*, Nature **455**, 78-80 (4 September 2008). arXiv:0809.2442

79 [133] **Composite photo.** Shin'ichiro Ando and Alexander Kusenko, *Evidence for Gamma-ray Halos Around Active Galactic Nuclei and the First*

Measurement of Intergalactic Magnetic Fields, 2010 ApJ 722 L39.

79 [134] **Composite photo reply**. Neronov, A., Semikoz, D.V., Tinyakov P.G., and Tkachev I.I., *No evidence for gamma-ray halos around active galactic nuclei resulting from intergalactic magnetic fields,* Astronomy and Astrophysics 06/2010; 526(11). arXiv:1006.0164v2

80 [136] **Stars S0-2 and S0-102**. Meyer, L.; Ghez, A. M.; Schödel, R.; Yelda, S.; Boehle, A.; Lu, J. R.; Do, T.; Morris, M. R.; Becklin, E. E.; Matthews, K., *The Shortest-Known-Period Star Orbiting Our Galaxy's Supermassive Black Hole. Science **338** (6103): 84–87 (2012).*

81 [137] **Stacy McGaugh interview.** Tam Hunt, *The System of the World: a dialogue with Prof. Stacy McGaugh,* https://tamhunt.medium.com/the-system-of-the-world-a-dialogue-with-prof-stacy-mcgaugh-fa1b3945f194

81 [137] **Synchronised satellite galaxies.** Oliver Müller, Marcel S. Pawlowski, Helmut Jerjen, Federico Lelli, *A whirling plane of satellite galaxies around Centaurus A challenges cold dark matter cosmology*, Science 02 Feb 2018: Vol. **359**, Issue 6375, pp. 534-537

Marcel S. Pawlowski, *The Planes of Satellite Galaxies Problem, Suggested Solutions, and Open Questions*, arXiv: 1802.02579 **[astro-ph.GA]**

81 [137] **DES.** Pallab Ghosh, *New dark matter map reveals cosmic mystery -* BBC News   https://www.bbc.co.uk/news/science-environment-57244708

82 [142] **Flyby anomaly.** P.G. Antreasian; J.R. Guinn (1998), *Investigations into the unexpected delta-v increase during the Earth Gravity Assist of GALILEO and NEAR*, AIAA/AAS Astrodynamics Specialist Conf. and Exhibition, Boston, paper no. 98-4287

82 [142] **Pioneer anomaly.** John D. Anderson et al., *Study of the anomalous acceleration of Pioneer 10 and 11*, Phys. Rev. D65:082004, 2002. arXiv:gr-qc/0104064v5

83 [143] **Jordan formula.** John D. Anderson et al., *Anomalous orbital-energy changes observed during spacecraft flybys of earth.* Phys Rev Lett. 100, 091102, 2008.

83 [144] **Busack.** H. J., *Expected velocity anomaly for the Earth flyby of Juno spacecraft on October 9, 2013,* arXiv:1312.1139

86 [149] **Hafele**. Joseph C., *Causal Version of Newtonian Theory by Time–*

Retardation of the Gravitational Field Explains the Flyby Anomalies,
Progress in Physics, Vol 2, April 2013.

86 [149] **2015 'Transversal component'.** L. Acedo, *The flyby anomaly: A case for strong gravitomagnetism?,* Advances in space research, 54(4), 788-796, arXiv:1505.06884 **[gr-qc]**

86 [149] **2017 'Unknown 5th force'.** Acedo, L. The flyby anomaly: *a multivariate analysis approach,* Astrophys Space Sci **362,** 42 (2017).

88 [152] **Expansion rate, new discrepancy**. Riess, Adam G et al., *A 2.4% Determination of the Local Value of the Hubble Constant, 2016,* arXiv:1604.01424

Measurement of Universe's expansion rate creates cosmological puzzle
http://www.nature.com/news/measurement-of-universe-s-expansion-rate-creates-cosmological-puzzle-1.19715?WT.mc_id=TWT_NatureNews

88 [152] **Standard candles**. Milne, Peter A, et al., *The changing fractions of type 1a supernova nuv-optical subclasses with redshift,* The Astrophysical Journal, Volume 803, Number 1, April 2015

Accelerating universe? Not so fast
http://www.eurekalert.org/pub_releases/2015-04/uoa-aun041015.php

88 [152] **Hubble tension.** Stuart Clark, *The universe is expanding too fast, and that could rewrite cosmology,* New Scientist Magazine, November 2020
https://www.newscientist.com/article/mg24833100-800

89 [154] **MOND.** M. Milgrom (2008), The MOND paradigm, arXiv:0801.3133

89 [155] **Dark matter and MOND.** McGaugh, S., *A Tale of Two Paradigms: the Mutual Incommensurability of LCDM and MOND,* Canadian Journal of Physics 93, 250 (2015) arXiv:1404.7525v2.

90 [155] **Lenses in clusters.** Massimo Meneghetti et al., *An excess of small-scale gravitational lenses observed in galaxy clusters,* Science 11 Sep 2020: Vol. 369, Issue 6509, pp. 1347-1351

Article. https://www.newscientist.com/article/2254246-dark-matter-in-galaxy-clusters-is-behaving-oddly-and-we-dont-know-why

90 [155] **Universe not clumpy enough.** http://kids.strw.leidenuniv.nl/KiDS-

1000.php , http://kids.strw.leidenuniv.nl/pr_jul2020.php

Article. https://www.newscientist.com/article/2250503-dark-matter-map-hints-at-cracks-in-our-understanding-of-the-universe

90 [155] **H2.** Valentijn, Edwin A.; van der Werf, Paul P., *First Extragalactic Direct Detection of Large-Scale Molecular Hydrogen in the Disk of NGC 891*, The Astrophysical Journal Lett, Volume 522, Issue 1, pp. L29-L33, Aug 1999

90 [156] **First cluster collision**. Douglas Clowe et al., *A direct empirical proof of the existence of dark matter,* Astrophys.J.648:L109-L113,2006, arXiv:astro-ph/0608407

90 [156] **Second cluster collision**. M. J. Jee, A. Mahdavi, H. Hoekstra, A. Babul, J. J. Dalcanton, P. Carroll, P. Capak, *A Study of the Dark Core in A520 with Hubble Space Telescope: The Mystery Deepens*, The Astrophysical Journal, Volume 747, Issue 2, article id. 96 (2012), arXiv:1202.6368

Article. http://hubblesite.org/newscenter/archive/releases/2012/10/full

90 [157] **Universal curve.** McGaugh, S. et al., *Radial Acceleration Relation in Rotationally Supported Galaxies*, Phys.Rev.Lett. 117 (2016) 20, 201101

90 [157] **Stacy McGaugh.** *Five laws of galactic rotation* [see pages 5 to 11], http://online.kitp.ucsb.edu/online/cdm-c18/mcgaugh/oh/05.html

92 [161] **Decaying dark matter.** Kanhaiya L. Pandey, Tanvi Karwal, Subinoy Das, *Alleviating the H_0 and σ_8 anomalies with a decaying dark matter model*, JCAP07(2020)026, arXiv:1902.10636v3

92 [161] https://www.newscientist.com/article/2332963-shredded-dwarf-galaxies-may-lack-dark-matter-to-hold-them-together

92 [161] **Giant structures.** https://www.vice.com/en/article/zmj7pw/theres-growing-evidence-that-the-universe-is-connected-by-giant-structures

95 [165] **Flyby, retardation.** Acedo, L. *Anomalous accelerations in spacecraft flybys of the Earth*. Astrophys Space Sci **362,** 225 (2017).

96 [167] **Earth-Sun distance**. Krasinsky, G. A.; Brumberg, V. A., *Secular increase of astronomical unit from analysis of the major planet motions, and its interpretation,* Celestial Mechanics & Dynamical Astronomy, Volume 90, Issue 3-4, pp. 267-288, 2004

101 [177] **Lorenzo Iorio**. *A-priori "imprinting" of General Relativity itself on some tests of it?* Advances in Astronomy Volume 2010 (2010), Article ID 735487. http://arxiv.org/abs/1002.4596

103 [179] **Dimensional quantum mechanics.** Kerr, J. M., *An interactions-based interpretation for quantum mechanics*, Physics essays, 33, 1 (2019)

103 [179] **DQM.** Kerr, J. M., *The Unsolved Puzzle*, Gordon Books, UK (2019). ISBN: 978-0956422262

103 [180] **RQM.** Rovelli, C., *"Relational Quantum Mechanics"*; International Journal of Theoretical Physics **35**; 1996: 1637-1678; arXiv:quant-ph/9609002

103 [180] **Kochen**. S. Symposium of the Foundations of Modern Physics: 50 Years of the Einstein-Podolsky-Rosen Gedanken experiment (World Scientific Publishing Co., Singapore, 1985), pp. 151–69.

103 [180] **Decoherence**. Schlosshauer, M., *Decoherence, the measurement problem, and interpretations of quantum mechanics.* Reviews of Modern Physics, 76 1267-1305 (2005), http://arxiv.org/abs/quant-ph/0312059

103 [181] **Entanglement quote.** R. M. Gingrich, A. J. Bergou, and C. Adami, *Entangled Light in Moving Frames*, Frontiers in Optics, OSA Technical Digest (CD) (Optical Society of America, 2003), paper WAA7.

103 [181] **Support for relational theories.** D. Frauchiger and R. Renner, *Quantum theory cannot consistently describe the use of itself*, Nature Communications 9(1), 3711 (2018)

103 [181] **Laboratory support for relational theories.** M. Proietti, A Pickston, F. Graffitti, P. Barrow, D. Kundys, C. Branciard, M. Ringbauer, A. Fedrizzi, *experimental rejection of observer-independence in the quantum world*, arXiv:1902.05080 (Submitted 13 Feb 2019).

105 [185] **Shapiro time delay.** Will, C. M., *Was Einstein Right?* Basic Books, New York, 1986, Oxford University Press 1995 edition, p117.

106 [187] **Clifford M. Will**. *The Confrontation between General Relativity and Experiment*, https://arxiv.org/abs/1403.7377 https://www.emis.de/journals/LRG/Articles/lrr-2006-3/fulltext.html

110 [198] **M. A. Abramowicz et al.,** *Leaving the innermost stable circular orbit: the inner edge of a black-hole accretion disk at various luminosities*, A&A, 521 (2010) A15, arXiv:1003.3887 [astro-ph.HE]

Appendix C [206] **Clifford M. Will**. *Finally, results from Gravity Probe B*, Physics 4, 43 (2011), arXiv:1106.1198

www.ingramcontent.com/pod-product-compliance
Lightning Source LLC
Chambersburg PA
CBHW060014210326
41520CB00009B/883